Ergebnisse der Mathematik und ihrer Grenzgebiete

Band 74

Herausgegeben von P. R. Halmos · P. J. Hilton
R. Remmert · B. Szőkefalvi-Nagy

Unter Mitwirkung von L. V. Ahlfors · R. Baer
F. L. Bauer · A. Dold · J. L. Doob · S. Eilenberg
K. W. Gruenberg · M. Kneser · G. H. Müller
M. M. Postnikov · B. Segre · E. Sperner

Geschäftsführender Herausgeber: P. J. Hilton

Robert M. Fossum

The Divisor Class Group of a Krull Domain

Springer-Verlag New York Heidelberg Berlin
1973

Robert M. Fossum, Professor of Mathematics
University of Illinois at Urbana-Champaign, Urbana, Illinois 61801, U.S.A.

AMS Subject Classifications (1970):

Primary   13D15, 13FXX, 14C20, 14M05

Secondary 13BXX, 13CXX, 13HXX, 14B15

ISBN 0-387-06044-8 Springer-Verlag New York Heidelberg Berlin
ISBN 3-540-06044-8 Springer-Verlag Berlin Heidelberg New York

To Jack

Table of Contents

Introduction

There are two main purposes for the writing of this monograph on factorial rings and the associated theory of the divisor class group of a Krull domain. One is to collect the material which has been published on the subject since Samuel's treatises from the early 1960's. Another is to present some of Claborn's work on Dedekind domains.

Since I am not an historian, I tread on thin ice when discussing these matters, but some historical comments are warranted in introducing this material. Krull's work on finite discrete principal orders originating in the early 1930's has had a great influence on ring theory in the succeeding decades. Mori, Nagata and others worked on the problems Krull suggested. But it seems to me that the theory becomes most useful after the notion of the divisor class group has been made functorial, and then related to other functorial concepts, for example, the Picard group. Thus, in treating the group of divisors and the divisor class group, I have tried to explain and exploit the functorial properties of these groups.

Perhaps the most striking example of the exploitation of this notion is seen in the works of I. Danilov which appeared in 1968 and 1970. Danilov studies the question: If A is factorial, is the ring of formal power series $A[[X]]$ factorial? He gives an almost complete answer to this question. But his techniques, algebro-geometric, rest on the geometric results of Grothendieck and Hironaka and use the functorial properties of the ideal class group introduced by Samuel (and Bourbaki).

A more elementary, but equally elegant, application of the techniques introduced by Samuel is made by Claborn to show that any abelian group is the ideal class group of some Dedekind domain. Since the Segre embedding $\mathbf{P}^1 \times \mathbf{P}^1 \to \mathbf{P}^3$ and its associated homogeneous coordinate ring is used to prove the theorem, one can say that geometry again makes its appearance in a purely algebraic setting. But even stronger connections are made by Claborn and Leedham-Green in their study of other properties of Dedekind domains. Of course one is led far afield from the classical rings of algebraic integers; the fundamental examples of Dedekind domains.

As a conclusion, these two theories, Danilov's for studying formal power series rings and Claborn's for studying Dedekind domains offer ample support for the study of the functorial properties of factorization theory.

In discussing these results, it is necessary to know the theory upon which they are based. Therefore I have included, in the first two chapters, a rapid development of the theory of Krull domains, the group of divisors and the divisor class group. One feature in this treatment is the inclusion of the theory of divisorial lattices, an extension of the theory of reflexive lattices over noetherian integrally closed integral domains. The results bear out the philosophy that divisorial lattices over a Krull domain look like torsion free finitely generated modules over a Dedekind domain. Moreover the theory has proved useful in other areas of mathematics.

Looking at the group of divisors of a Krull domain as the set of integer valued functions with finite support defined on the set of minimal nonzero prime ideals permits the precise statement of the approximation theorem and subsequently a good formulation of Claborn's existence problem as well as the generalization later to schemes.

Among the more important results which appear in these first two chapters are the Theorem of Mori-Nagata, which states that the integral closure of a noetherian domain is a Krull domain, Nagata's Localization Theorem for the ideal class group, Auslander and Buchsbaum's result concerning the factoriality of regular local rings and Murthy's Theorem which states that a Cohen-Macaulay factorial local ring is Gorenstein.

A good deal of space is devoted to using Nagata's Localization Theorem, principally in studying graded rings. The reason for this is that many examples of factorial rings can be obtained. Even some excercises in Bourbaki are solved because the examples are useful.

In the third chapter we discuss Claborn's results concerning Dedekind domains. His work on Dedekind domains can be said to have originated from a problem in Zariski and Samuel's first volume on Commutative Algebra: Is every Dedekind domain the integral closure of a principal ideal domain? He was able to show that this is far from the case. But his investigations led to many interesting results. The first one mentioned here is his generalized approximation theorem which is a corollary of the theorem: If the cardinality of the Dedekind domain A is larger than (card Spec $A)^{\aleph_0}$, then A is a principal ideal domain. The second result is that a given abelian group is isomorphic to the ideal class group of some Dedekind domain. These two results are discussed in the first two sections of Chapter III.

The last section of this chapter is devoted to exposing the results in an unpublished manuscript of Claborn's. These results generalize to the uncountable case his published work concerning the construction of Dedekind domains whose maximal ideals are in one-to-one correspondence with the generators of a free abelian group and whose principal ideals generate a given subgroup. There are examples, in the uncountable case, of a group and a subgroup which do not admit a Dedekind domain with these desired properties and a Dedekind domain whose group of principal ideals is not dense in the group of divisors. The basic reasons for the general construction are realized in the countable case. However in view of the fact that every abelian group is isomorphic to the divisor class group of some Dedekind domain, there is interest in the uncountable case. The order and notation of Claborn's manuscript have been changed somewhat, but the essential points of the proofs have not been changed.

In Chapter IV we reproduce Samuel's theory of descent, which is most useful for constructing examples as demonstrated by Bertin. She uses the theory to find a factorial ring which is not Cohen-Macaulay. Examples also arise which show that power series adjunction and completion do not preserve factoriality. Brief mention is made, in an appendix to the chapter, of Waterhouse's generalization which combines the two theories of Galois and radical descent by using the theory of Hopf algebras.

In the final chapter the relations between the Picard group of a ringed space and the divisor class group are established and used. The properties of the Picard group and its relations to the divisor class group are established in the first section (§ 18). One of the most interesting applications is Danilov's result which shows the existence of a natural splitting for the monomorphism $\mathrm{Cl}(A) \to \mathrm{Cl}(A[[X]])$. The final section is devoted to a discussion of Danilov's results which say when the homomorphism $\mathrm{Cl}(A) \to \mathrm{Cl}(A[[X]])$ is a bijection. We have included Grothendieck's theorem which states that a complete intersection which is factorial in codimension 3 is factorial. Here is an example of a purely algebraic theorem proved with most sophisticated techniques of algebraic geometry. One of the open problems is to give a purely algebraic proof of this result. In discussing this result and Danilov's results we must omit the proofs, since they involve results outside the scope of this treatise. A most recent paper on this subject by [Danilov, 1972] appeared too late for inclusion in this last section.

A few comments are in order concerning the contents and completeness of the text. In addition to the unpublished work of Claborn's in Chapter III there are new results sprinkled throughout. In particular some of the material about divisorial lattices has not been published before. Also the material concerning graded noetherian Krull domains following Samuel's treatment has been extended. And there have been attempts at providing concise proofs within the context of the book. However, in the final two chapters especially, no attempt has been made to give complete details of the demonstrations of the results since many deep theorems from algebraic geometry would be needed.

Several interesting topics have been omitted partly because they do not fit into the functorial picture, partly because of my own prejudices. Among these which are important are: Goldman's construction of Dedekind domains, the general theory of Euclidean rings and the results concerning curves whose coordinate rings are Euclidean, and the many results concerning power series rings in infinitely many variables. Finally, only brief mention is made of many other very interesting results.

The list of references, not all used in the text, includes hopefully all the important papers since approximately 1960. However, no claim is made that it is exhaustive.

A few peculiarities need mention. References to the literature are made by name of author and year of publication, e.g., [Halmos, 1971]. The papers of an author which appear in the same year are denoted by a, b, c, etc., e.g., [Danilov, 1970a]. Internal references are made by section number and result number, e.g., Proposition 20.1 is the first result in Section 20. Upper case letters from the end of the alphabet which appear inside brackets, as for example $A[[X]]$, always mean indeterminates. A few terms and notations which I feel need mention are included in a short appendix after Section 19. In other cases, definitions are included in the text. Otherwise I have adhered closely to basic notation in commutative ring theory.

Originally this book was planned as a joint work with Luther Claborn. Unfortunately he was fatally injured in an automobile accident in August, 1967, shortly after a basic outline of the book had been approved by the editors of this series. His wife, Mary Lee Claborn, was most helpful in finding the manuscript which contains the details of Claborn's work reported in Section 15. I take this opportunity to thank her for her helpfulness and encouragement in completing this project.

Professor Jack E. McLaughlin taught Claborn and me at The University of Michigan. I am grateful to him for his inspiration and guidance during the preparation of this book. I also take this opportunity to thank Professor Irving Kaplansky for reading a version of the manuscript and Professor Peter Hilton, who, as an editor of this series, showed continued interest and made helpful suggestions. I am appreciative of the assistance and cooperation of Dr. Klaus Peters and his coworkers at Springer-Verlag. And I am especially indebted to the University of Illinois at Urbana-Champaign, Universitetet i Oslo, Aarhus Universitet,

The United States Educational Foundation in Norway and the United States National Science Foundation for generous financial support at various times during the preparation of this monograph.

Aarhus, Denmark   Robert M. Fossum
June, 1972

Chapter I. Krull Domains

The purpose of this chapter is to introduce and develop the notion of a Krull domain. Since most of the material appears in readily accessible sources, no effort is made to give complete demonstrations of the statements. The chapter is divided into five sections.

Section 1 is devoted to the definition of a Krull domain and the statements concerning the closure of the class of Krull domains under certain ring extensions (for example the morphism $A \to A[X]$).

Section 2 and Section 5 are devoted to the discussion of lattices over Krull rings, and to the theory of divisorial ideals.

In Section 3 we give a treatment of the relation which exists between Krull domains and completely integrally closed integral domains.

In Section 4 we present Krull's normality conditions for a noetherian domain and the Mori-Nagata Theorem which states that the integral closure of a noetherian integral domain is a Krull domain.

The importance of Krull rings in the theory of commutative rings rests on, among others, the results:

The divisors form a free abelian group based on the height one prime ideals.

The integral closure of a noetherian integral domain is a Krull ring.

A unique factorization domain is a Krull ring.

§ 1. The Definition of a Krull Ring

Let A be an integral domain which is contained in a field K. The integral domain A is said to be a *Krull* domain provided there is a family $\{V_i\}_{i \in I}$ of principal valuation rings (i.e., discrete rank one valuation rings), with $V_i \subseteq K$, such that

(I)
$$A = \bigcap_{i \in I} V_i$$

(FC) Given f in A, $f \neq 0$, there is at most a finite number of i in I such that f is not a unit in V_i.

The first property (I) is called the *intersection* property. The property (FC) is called the *finite character* property for the collection of rings $\{V_i\}_{i \in I}$.

Krull [1931] called such rings finite discrete principal orders. The theory of Krull domains, in particular the proofs of the next few statements, appears in numerous references, among them Krull [loc. cit.], Zariski and Samuel [1960], Nagata [1962], Samuel [1964d], Bourbaki [1965], and Gilmer [1968].

In the following, it is assumed that A is a Krull domain with field of quotients K.

Proposition 1.1. *A Krull domain is integrally closed.*   □

Proposition 1.2. *If k is a subfield of K, then $A \cap k$ is a Krull domain.*   □

Proposition 1.3. *If L is a finite extension field of K, then the integral closure of A in L is a Krull domain.*   □

Proposition 1.4. *If $\{A_i\}_{i \in I}$ is a family of Krull domains each contained in K and if the family satisfies the finite character property, then $\bigcap_{i \in I} A_i$ is a Krull domain.*   □

Corollary 1.5. *If $A = \bigcap_{i \in I} V_i$ and if J is a subset of I, then $\bigcap_{j \in J} V_j$ is a Krull domain.*   □

The Krull domain $\bigcap_{j \in J} V_j$ is called a *subintersection* of A.

Proposition 1.6. *If $\{T_i\}_{i \in J}$ is a family of indeterminates, then the polynomial ring $A[\{T_i\}_{i \in J}]$ is a Krull domain.*   □

Proposition 1.7. *If T is an indeterminate, then the ring of formal power series $A[[T]]$ is a Krull domain.*   □

There are at least three distinct rings of formal power series in infinitely many indeterminates. These are all Krull rings in case the coefficient ring is a Krull domain. See Gilmer's interesting paper about these three rings [Gilmer, 1969].

If A is a ring, then A^* will denote the units of A.

Proposition 1.8. *Suppose $A = \bigcap_{i \in I} V_i$ and suppose S is a multiplicatively closed subset of A. Then the ring of quotients $S^{-1}A = \bigcap_{j \in J} V_j$, where $J = \{i \in I : S \subseteq V_i^*\}$, is a Krull domain.*   □

Recall that a valuation overring V of a ring A is essential provided $V = A_{\mathfrak{q}}$ for some prime ideal $\mathfrak{q}$ in Spec A. Using the essential principal valuation overrings of the Krull domain A we can obtain the most efficient representation of A as the intersection of principal valuation rings.

Let $X^{(1)}(A)$ (or just $X^{(1)}$) denote the prime ideals of height one in A.

Proposition 1.9. *The integral domain A is a Krull domain if and only if the three conditions below are satisfied.*

(a) *For each $\mathfrak{p}$ in $X^{(1)}(A)$, the local ring $A_{\mathfrak{p}}$ is a principal valuation ring.*

(b) $A = \bigcap_{\mathfrak{p} \in X^{(1)}} A_{\mathfrak{p}}.$

(c) *Each f in A, $f \neq 0$, is contained in at most a finite number of prime ideals of height one.* $\Box$

Condition (c) is the finite character property for the family $\{A_{\mathfrak{p}}\}_{\mathfrak{p} \in X^{(1)}(A)}$. This proposition will be demonstrated in Section 3.

It is clear from Proposition 1.6 that there are Krull domains which are not noetherian. However, what is not clear is whether there are Krull rings with a height one prime ideal which is not of finite type. Eakin and Heinzer [1970] have provided an example.

Example 1.10. Let k be a field and set $B = k[X_1, \ldots]$, the ring of polynomials in countably many variables $\{X_i\}$. Let A be the subring of B generated over k by the products $X_i X_j$ for all pairs i, j. Then A is a Krull domain. Set $\mathfrak{P} = X_1 B$, a prime ideal of height one in B. Then $\mathfrak{P} \cap A = \sum_i X_1 X_i A$, and since B is integral over A and the height of $\mathfrak{P}$ is one, $\mathrm{ht}\, \mathfrak{P} \cap A = 1$. The prime ideal $\mathfrak{P} \cap A$ cannot be generated by a finite number of elements. For each element in A involves only a finite number of variables. Each element f in $\mathfrak{P} \cap A$ is a product $X_1 g$ for g in B. Let $f_1, \ldots, f_r$ be a set of generators of $\mathfrak{P} \cap A$, and write each $f_i = X_1 g_i$. Let X_N be a variable not involved in any of the f_i. Then

$$X_1 X_N = \sum_1^r a_i f_i = X_1 \sum_1^r a_i g_i. \quad \text{So} \quad X_N = \sum_i a_i g_i \quad \text{in } B.$$ Since each g_j is in

the maximal ideal of B generated by all the variables, and none involves X_N, this equation leads to a contradiction. Thus $\mathfrak{P} \cap A$ is a prime ideal of height one which is not finitely generated.

§ 2. Lattices

Suppose that A is an integral domain with field of quotients K. Let V be a finite dimensional vector space over K. An A-submodule M of V is an *A-lattice* in V provided

(i) M contains a basis for V as a K-module (i.e., $KM = V$) and M is a submodule of a finitely generated A-submodule of V.

Proposition 2.1. *The following conditions are equivalent to* (i) *above for an A-submodule of the vector space V.*

(ii) *There are free A-modules F, F' in V of the same rank as the dimension of V such that $F \subseteq M \subseteq F'$.*

(iii) *There is a free A-submodule $F \subseteq V$ of the same rank as the dimension of V and an $f \neq 0$, $f \in A$, such that $fF \subseteq M \subseteq F$.*

(iv) *There are A-lattices N, N' in V such that $N' \subseteq M \subseteq N$.*

(v) *Given any free A-submodule F of V of the same rank as the dimension of V, there are f, g in A, $fg \neq 0$, such that $fF \subseteq M \subseteq g^{-1}F$.* $\square$

The next proposition, almost word for word the same as in Bourbaki [1965], is useful in constructing new A-lattices from given A-lattices, and is fundamental for that reason.

Proposition 2.2. *Let $U, V, V_1, \ldots, V_r, W$ be finite dimensional vector spaces over K.*

(i) *If M, N are A-lattices in V, then $M + N$ and $M \cap N$ are A-lattices in V.*

(ii) *If $U \subseteq V$ and M is an A-lattice in V, then $M \cap U$ is an A-lattice in U.*

(iii) *If $\mu : V_1 \times \cdots \times V_r \to U$ is a multilinear form and if $M_1, \ldots, M_r$ are A-lattices in the $V_1, \ldots, V_r$ respectively, then the A-module generated by $\mu(M_1 \times \cdots \times M_r)$ is an A-lattice in the subspace spanned by $\mu(V_1 \times \cdots \times V_r)$.*

(iv) *Let M be an A-lattice in V, N an A-lattice in W. Then $N : M = \{\alpha \in \mathrm{Hom}_K(V, W) : \alpha(M) \subseteq N\}$ is an A-lattice in $\mathrm{Hom}_K(V, W)$.*

(v) *If B is an integral domain with field of quotients L where L is an extension field of K such that $A \subseteq B$, and if M is an A-lattice in V, then $BM = \mathrm{Im}(B \otimes_A M \to L \otimes_K V)$ is a B-lattice in $L \otimes_K V$.*

Proof. (i) Let F be a free A-submodule of V with rank $F = \dim V$. Since $M \cap N \subseteq M + N$ it suffices to find a, b in A such that $aF \subseteq M \cap N$ and $b^{-1}F \supseteq M + N$. Suppose that $f_1 F \subseteq M$ and $g_1 F \subseteq N$. Then $f_1 g_1 F \subseteq M \cap N$. Now suppose $M \subseteq f^{-1}F$ and $N \subseteq g^{-1}F$. Then $M + N \subseteq (fg)^{-1}F$.

(ii) Let $u_1, \ldots, u_r$ be a basis for U. Extend this to a basis $u_1, \ldots, u_n$ for V. Let G be the free A-submodule of U spanned by $u_1, \ldots, u_r$ and F the free A-submodule of V spanned by all the u_j. Then $F \cap U = G$. Further, if $f \in A$, then $f(F \cap U) = (fF) \cap U$. Now if M is an A-lattice in V, then there are f, g in A, $fg \neq 0$, such that $fF \subseteq M \subseteq g^{-1}F$ and so $fG \subseteq M \cap U \subseteq g^{-1}G$. Hence $M \cap U$ is an A-lattice in U.

(iii) To see this statement, it is sufficient to show that the vector spaces spanned over K by $\mu(M_1 \times \cdots \times M_r)$ and $K\mu(V_1 \times \cdots \times V_r)$ are equal and that $\mu(M_1 \times \cdots \times M_r)$ is contained in a finitely generated A-submodule of the second. But both these follow from the corresponding statements for the M_j.

(iv) Let F, G be free A-lattices in V, W respectively. Consider the natural homomorphism $\operatorname{Hom}_A(F, G) \to \operatorname{Hom}_K(V, W)$ obtained by the composition of $\operatorname{Hom}_A(F, G) \to K \otimes_A \operatorname{Hom}_A(F, G)$ with the identification of the second group with $\operatorname{Hom}_K(V, W)$. Since F is free of finite type and G is torsion free, this homomorphism is an injection. Its image is contained in $G:F$. On the other hand, any element of $G:F$ can be considered to be an element of $\operatorname{Hom}_A(F, G)$ by restriction. So the two groups are identified by this injection. Thus $G:F$ is a lattice in $\operatorname{Hom}_K(V, W)$ and it is a free A-module. The following lemma, whose proof is obvious, is needed now. But it will be useful later as well.

Lemma 2.3. *Let $M' \subseteq M$ and $N \subseteq N'$ be A-lattices in V and W respectively. Then $N:M \subseteq N':M'$.* $\square$

With this lemma in mind, we return to the proof of (iv). Suppose that F, F', M are lattices in V and G, G', N are lattices in W with F, F', G, G' free. If $F \subseteq M \subseteq F'$ and $G' \subseteq N \subseteq G$, then $G:F \supseteq N:M \supseteq G':F'$ and thus $N:M$ is an A-lattice in $\operatorname{Hom}_K(V, W)$.

It has been seen that $G:F$ can be identified with $\operatorname{Hom}_A(F, G)$ when F and G are free. This is true quite in general, for restriction induces an isomorphism $N:M \to \operatorname{Hom}_A(M, N)$.

(v) If $F \subseteq M \subseteq F'$, then $BF \subseteq BM \subseteq BF'$. But BF and BF' are both of finite type as B-modules if F and F' are of finite type as A-modules. Also, if F contains a basis for V, then BF contains a basis for $L \otimes_K V$. Thus BM is a B-lattice in $L \otimes_K V$. $\square$

There is special interest in the A-lattices of the form $A:(A:M)$ where M is an A-lattice in V. Note first, that $\operatorname{Hom}_K(\operatorname{Hom}_K(V, K), K)$ is naturally identified with V by the homomorphism $c_V : V \to \operatorname{Hom}_K(\operatorname{Hom}_K(V, K), K)$ defined by $c_V(v)(\alpha) = \alpha(v)$ for v in V and α in $\operatorname{Hom}_K(V, K)$. Now M is an A-submodule of $A:(A:M)$. For if $m \in M$ and $\alpha \in A:M$, then $c_V(m)(\alpha) \in A$. Furthermore, if F is a free A-lattice in V, then $F = A:(A:F)$ (with the identification c_V always kept in mind). To see this, let $e_1, \dots, e_n$ be an A-basis for F. Let $e_1^*, \dots, e_n^*$ be the basis of $\operatorname{Hom}_K(V, K)$ dual to the basis of F. Then $\alpha \in A:F$ if and only if $\alpha \in \sum A e_i^*$. Then an element v in V is in $A:(A:F)$ if and only if $v \in \sum A e_i^{**} = \sum A e_i = F$.

The next result shows that this equality holds for a class of A-lattices much larger than the free ones.

Lemma 2.4. *Let M be an A-lattice in V. Then $A:M = A:(A:(A:M))$.*

Proof. It has been seen that $A:M \subseteq A:(A:(A:M))$. On the other hand, since $M \subseteq A:(A:M)$, it follows from the previous lemma, that $A:M \supseteq A:(A:(A:M))$. $\square$

An A-lattice M for which $M = A:(A:M)$ is called a *divisorial* A-lattice. The last lemma shows that a lattice M is divisorial if and only if $M = A:N$ for some A-lattice N. Some fundamental properties of divisorial A-lattices are now given. They are very close to the general properties given above for lattices, and in fact they show when the operations preserve the property of being divisorial.

Proposition 2.5. *Let* M *be an* A-lattice *in* V. *Then* $A:(A:M) = \bigcap_{\alpha \in A:M} (\alpha^{-1}(A) \cap F)$ *where* F *is a free* A-lattice *containing* M. *And so* M *is divisorial if and only if* $M = \bigcap_{\alpha \in A:M} (\alpha^{-1}(A) \cap F)$.

Proof. Let α be in $A:M$. Then $M \subseteq \alpha^{-1}(A) \cap F \subseteq F$. In particular $\alpha^{-1}(A) \cap F$ is an A-lattice. Furthermore, it is divisorial. For suppose $v \in A:(A:(\alpha^{-1}(A) \cap F))$. Now $\alpha \in A:(\alpha^{-1}(A) \cap F)$. Hence $\alpha(v) \in A$ which implies $v \in \alpha^{-1}(A)$. But F itself is divisorial, so $A:(A:(\alpha^{-1}(A) \cap F)) \subseteq F$. The conclusion is $A:(A:(\alpha^{-1}(A) \cap F)) \subseteq \alpha^{-1}(A) \cap F$ and thus $\alpha^{-1}(A) \cap F$ is divisorial. Hence $A:(A:M)$ is contained in the intersection.

Suppose, in order to prove the opposite containment, that v is an element in the intersection. To show that $v \in A:(A:M)$ it is sufficient to show that $\alpha(v) \in A$ for all α in $A:M$. But for such an α, $v \in \alpha^{-1}(A) \cap F$, and hence the desired conclusion follows. $\square$

Proposition 2.6. *Let* N *be an* A-lattice (*in some vector space*) *and* M *a divisorial* A-lattice. *Then* $M:N$ *is a divisorial* A-lattice.

The proof of this result needs some functorial properties of the operation $M:N$.

Lemma 2.7. *Let* M, N, L *be* A-lattices *in the vector spaces* U, V, W *respectively. Let* adj: $\mathrm{Hom}_K(U \otimes_K V, W) \to \mathrm{Hom}_K(U, \mathrm{Hom}_K(V, W))$ *be the canonical adjoint isomorphism. Let* MN *denote the image of* $M \otimes_A N$ *in* $U \otimes_K V$. *Then*
  (i) adj$(L:MN) \subseteq (L:N):M$
  (ii) adj$^{-1}((L:N):M) \subseteq L:MN$.
  Let $\sigma: \mathrm{Hom}_K(V, W) \otimes_K U \to \mathrm{Hom}_K(\mathrm{Hom}_K(U, V), W)$ *be defined by* $\sigma(\alpha \otimes u)(\beta) = \alpha(\beta(u))$. *Then* σ *is an isomorphism and*
  (iii) $\sigma((L:N)M) \subseteq L:(N:M)$.

Proof. Let $\alpha(MN) \subseteq L$ for some α. Then, for all $m \in M$, it follows that adj$(\alpha)(m)(n) = \alpha(m \otimes n) \in L$. Hence adj$(\alpha)M \subseteq L:N$ and this proves that adj$(\alpha) \in (L:N):M$. The second statement follows similarly.

In order to prove the third, let $\beta \in N:M$, $\alpha \in L:N$ and $m \in M$. Then it follows that $\sigma(\alpha \otimes m)(\beta) = \alpha(\beta(m)) \in L$. Hence $\sigma(\alpha \otimes m)(N:M) \subseteq L$ and the containment is proved. $\square$

Now it is easy to establish the proposition. Since M is divisorial $M = A:(A:M)$. Hence $M:N = (A:(A:M)):N$. By the first statement in the lemma $M:N \supseteq A:(A:M)N$. By the third statement $(A:M)N \subseteq A:(M:N)$ and hence $A:(A:M)N \supseteq A:(A:(M:N))$. So finally $M:N \supseteq A:(A:(M:N)) \supseteq M:N$ and the proposition is established. (Both adj and σ have been considered to be identifications in this proof.) $\square$

A *fractionary ideal* is an A-lattice in K. The results in the above section apply to fractionary ideals. Now an element in $\text{Hom}_K(K,K)$ is multiplication by an element in K. It follows that $A:\mathfrak{a}$ is a fractionary ideal whenever $\mathfrak{a}$ is a fractionary ideal. Proposition 2.5 becomes:
$$A:(A:\mathfrak{a}) = \bigcap_{x\mathfrak{a} \subseteq A} x^{-1}A = \bigcap_{\mathfrak{a} \subseteq yA} yA.$$ Thus $\mathfrak{a}$ is divisorial if and only if $\mathfrak{a}$
is the intersection of principal fractionary ideals. Proposition 2.6 becomes: Let $\mathfrak{a}, \mathfrak{b}$ be fractionary ideals. If $\mathfrak{a}$ is divisorial, then $\mathfrak{a}:\mathfrak{b} = \{x \in K : x\mathfrak{b} \subseteq \mathfrak{a}\}$ is a divisorial ideal. Lemma 2.7 says: $\mathfrak{a}:\mathfrak{b}\mathfrak{c} = (\mathfrak{a}:\mathfrak{b}):\mathfrak{c}$ and $(\mathfrak{a}:\mathfrak{b})\mathfrak{c} \subseteq \mathfrak{a}:(\mathfrak{b}:\mathfrak{c})$ for the fractionary ideals $\mathfrak{a}, \mathfrak{b}, \mathfrak{c}$. As an example, it follows that $A:((A:\mathfrak{a})\mathfrak{a}) = (A:\mathfrak{a}):(A:\mathfrak{a})$. Denote $A:\mathfrak{a}$ by $\mathfrak{a}^*$.

§ 3. Completely Integrally Closed Rings

Let x be an element in the field of quotients of A. The element x is said to be *almost integral* over A if the ideal $\sum_{i=0}^{\infty} A x^i$ is a fractionary ideal. This is the same as to say that the powers of x have a common denominator.

Lemma 3.1. *An element x is almost integral over A if and only if there is a fractionary ideal $\mathfrak{a}$ such that $x\mathfrak{a} \subseteq \mathfrak{a}$.*

Proof. If x is almost integral over A, then $\mathfrak{a} = \sum_{i=0}^{\infty} A x^i$ is a fractionary ideal with the desired property. On the other hand, if $x\mathfrak{a} \subseteq \mathfrak{a}$, then $x^i \in \mathfrak{a}:\mathfrak{a}$ which is a fractionary ideal. Thus the powers of x generate a fractionary ideal and x is almost integral over A. $\square$

Corollary 3.2. *The set $\tilde{A}$ of elements almost integral over A forms a ring.*

Proof. Suppose x, y are almost integral over A, say $x \in \mathfrak{a}:\mathfrak{a}$ and $y \in \mathfrak{b}:\mathfrak{b}$ for fractionary ideals $\mathfrak{a}$ and $\mathfrak{b}$. Then $(x+y) \in (\mathfrak{a} \cap \mathfrak{b}:\mathfrak{a} \cap \mathfrak{b})$ and

$xy \in \mathfrak{a}\mathfrak{b}:\mathfrak{a}\mathfrak{b}$. But $\mathfrak{a} \cap \mathfrak{b}$ and $\mathfrak{a}\mathfrak{b}$ are again fractional ideals, so $x+y$ and xy are again almost integral over A. This shows that $\tilde{A}$ is a ring. $\square$

Caution: It may happen that $\tilde{\tilde{A}} \neq \tilde{A}$ (cf. [Gilmer and Heinzer, 1968]).

The ring $\tilde{A}$ is called the *complete integral closure* of A and A is said to be completely integrally closed if $A = \tilde{A}$.

Corollary 3.3. *The integral domain A is completely integrally closed if and only if $A = \mathfrak{a}:\mathfrak{a}$ for all fractional ideals $\mathfrak{a}$.* $\square$

Let $D(A)$ denote the collection of divisorial fractional A-ideals. Define $D(A) \times D(A) \to D(A)$ by $(a,b) \mapsto A:(A:ab)$. Using the calculus of the operation $A:\mathfrak{a}$, it is seen that this is an associative operation which admits A as an identity. Hence $D(A)$ is a commutative monoid.

Proposition 3.4. *The monoid $D(A)$ is a group if and only if A is completely integrally closed.*

Proof. Suppose, on the one hand, that $D(A)$ is a group. Let $\mathfrak{a}$ be a fractional ideal. Suppose $\mathfrak{c} = \mathfrak{a}:\mathfrak{a}$. Let $\mathfrak{b} = A:(A:\mathfrak{a})$. If $x \in \mathfrak{c}$ and $y\mathfrak{a} \subseteq A$, then $xy\mathfrak{a} \subseteq A$. Hence x in $\mathfrak{c}$ implies $x \in (A:\mathfrak{a}):(A:\mathfrak{a})$. So $\mathfrak{c} \subseteq \mathfrak{b}:\mathfrak{b}$. Let $B = \mathfrak{b}:\mathfrak{b}$ be in $D(A)$. Then the product, in $D(A)$, of B with itself is $A:(A:B^2)$. But B is a ring, so $B^2 = B$. Hence $A:(A:B^2) = A:(A:B) = B$. Hence B is an idempotent in the group $D(A)$ so B is the identity element A. Hence $A \subseteq \mathfrak{a}:\mathfrak{a} \subseteq B = A$ so A is completely integrally closed.

If, on the other hand, A is completely integrally closed and $\mathfrak{a}$ is a divisorial fractional ideal, then $A:\mathfrak{a}(A:\mathfrak{a}) = (A:\mathfrak{a}):(A:\mathfrak{a}) = A$. Hence $A:(A:\mathfrak{a}(A:\mathfrak{a})) = A$ which shows that $A:\mathfrak{a}$ is an inverse to $\mathfrak{a}$ in $D(A)$. So $D(A)$ is a group. $\square$

Remark. $D(A)$ has a structure as a partially ordered monoid with $\mathfrak{a} \geqslant \mathfrak{b}$ provided $A:\mathfrak{a} \subseteq A:\mathfrak{b}$. If $\mathfrak{a}$ is a minimal non-identity element in $D(A)$, then $A:(A:\mathfrak{a})$ is a maximal proper divisorial ideal of A. This leads to the next two results.

Proposition 3.5. *Suppose A is completely integrally closed. Let $\mathfrak{a}$ be a maximal proper divisorial ideal in A. Then there is an x in K such that $\mathfrak{a} = A:(Ax+A)$ and $\mathfrak{a}$ is a prime ideal.*

Proof. Since $\mathfrak{a}$ is divisorial, $\mathfrak{a} = A:(A:\mathfrak{a})$ and since $\mathfrak{a}$ is proper, $A:\mathfrak{a} \neq A$. Let x be in $A:\mathfrak{a}$, $x \notin A$. Then $A \subsetneqq Ax+A \subseteq A:\mathfrak{a}$, so $\mathfrak{a} = A:(A:\mathfrak{a}) \subseteq A:(Ax+A) \subsetneqq A$. Since $\mathfrak{a}$ is maximal $\mathfrak{a} = A:(Ax+A)$.

Now suppose $\mathfrak{b}$ and $\mathfrak{c}$ are ideals in A such that $\mathfrak{b}\mathfrak{c} \subseteq \mathfrak{a}$. It can be assumed that $\mathfrak{a} \subseteq \mathfrak{c}$ and $\mathfrak{a} \subseteq \mathfrak{b}$. The object is to show that either $\mathfrak{b}$ or $\mathfrak{c}$ is contained in $\mathfrak{a}$. Suppose $\mathfrak{c}$ is not contained in $\mathfrak{a}$. Then $\mathfrak{a} \subsetneqq \mathfrak{c} \subseteq \mathfrak{a}:\mathfrak{b}$ $\subseteq \mathfrak{a}:\mathfrak{a} = A$. But $\mathfrak{a}:\mathfrak{b}$ is a divisorial ideal properly containing $\mathfrak{a}$. Hence $\mathfrak{a}:\mathfrak{b} = A$, and so $\mathfrak{b} \subseteq \mathfrak{a}$. $\square$

Theorem 3.6. *If A is a Krull domain, then:*
(i) *A is completely integrally closed.*
(ii) *The set of divisorial ideals satisfies the ascending chain condition.*

Proof. Suppose A is a Krull domain. Then $A = \bigcap V$ where the V are principal valuation rings. Let $\mathfrak{a}$ be a nonzero ideal. Then $\mathfrak{a}V$ is a nonzero principal ideal in V. If x in K is such that $x\mathfrak{a} \subseteq \mathfrak{a}$, then $x\mathfrak{a}V \subseteq \mathfrak{a}V$. But then $xV \subseteq V$, so $x \in V$. ($\mathfrak{a}V = tV$ for some t in $\mathfrak{a}$. Thus $x\mathfrak{a}V \subseteq \mathfrak{a}V$ if and only if $xtV \subseteq tV$ and this is the case if and only if $xV \subseteq V$.) Hence $x \in \bigcap V = A$ so $A = \mathfrak{a}:\mathfrak{a}$. So $A = \mathfrak{a}:\mathfrak{a}$ for all fractional ideals, whence A is completely integrally closed. (This argument shows that the intersection of principal valuation rings is completely integrally closed. The converse is not true.)

Let $\mathfrak{a}$ be an ideal in A. Consider $\bigcap \mathfrak{a}V$. Certainly $\mathfrak{a} \subseteq \bigcap \mathfrak{a}V$. If $\mathfrak{a} = (0)$, then $\mathfrak{a} = \bigcap \mathfrak{a}V$. Suppose $\mathfrak{a} \neq (0)$. Then $\mathfrak{a}V = V$ for all but a finite number of the V, since there is a nonzero f in $\mathfrak{a}$ which is a unit in almost all the V. If $\mathfrak{c}$ is a fractional A-ideal, then $\mathfrak{c}V$ is a fractional V-ideal for each V. In order to complete the proof, it is desirable to have a relation between $A:\mathfrak{c}$ and the $V:\mathfrak{c}V$. In fact

$$A:\mathfrak{c} = \bigcap (V:\mathfrak{c}V).$$

For if $x \in A:\mathfrak{c}$, then $x\mathfrak{c}V \subseteq V$ for all V, so $x \in \bigcap (V:\mathfrak{c}V)$. On the other hand $\mathfrak{c} \subseteq \mathfrak{c}V$, so x in $\bigcap (V:\mathfrak{c}V)$ gives $x\mathfrak{c} \subseteq V$ for all V and hence $x\mathfrak{c} \subseteq \bigcap V = A$. In particular, if $\mathfrak{c}$ is a divisorial fractional ideal, then $\mathfrak{c} = \bigcap (V:(A:\mathfrak{c})V)$.

Now let $(0) \neq \mathfrak{a}_1 \subseteq \mathfrak{a}_2 \subseteq \cdots$ be an ascending chain of divisorial ideals. Then $V:(A:\mathfrak{a}_1)V \subseteq V:(A:\mathfrak{a}_2)V \subseteq \cdots$ is an ascending chain of ideals in V for each V. Also $(A:\mathfrak{a}_1)V = V$ for almost all V. For the other finite number of principal valuation rings, there is an n such that $V:(A:\mathfrak{a}_n)V = V:(A:\mathfrak{a}_m)V$ for all $m \geq n$ (since each V is noetherian). Hence $V:(A:\mathfrak{a}_n)V = V:(A:\mathfrak{a}_m)V$ for $m \geq n$ and for all V. Hence $\mathfrak{a}_m = \bigcap (V:(A:\mathfrak{a}_m)V) = \bigcap (V:(A:\mathfrak{a}_n)V) = \mathfrak{a}_n$ for all $m \geq n$ so the chain stops. $\square$

The proof of the next theorem requires a few working lemmata which are now given. This approach to Theorem 3.10 is due to Istvan Beck.

Lemma 3.7. *Let A be an integral domain. Suppose $\mathfrak{p} \supsetneq \mathfrak{q}$ are ideals with $\mathfrak{q}$ prime. Then $A:\mathfrak{p} \subseteq \mathfrak{q}:\mathfrak{q}$.*

Proof. Let f be in $\mathfrak{p} - \mathfrak{q}$. Then $(A:\mathfrak{p})f \subseteq A$ so $(A:\mathfrak{p})f\mathfrak{q} \subseteq \mathfrak{q}$. As $f \notin \mathfrak{q}$, we get $(A:\mathfrak{p})\mathfrak{q} \subseteq \mathfrak{q}$, which gives the result. $\square$

Lemma 3.8. *Let $\mathfrak{p}$ be a prime ideal of A. Then $A_{\mathfrak{p}} = \{x \in K^*: Ax^{-1} \cap A \not\subseteq \mathfrak{p}\} \cup \{0\}$.*

Proof. If $0 \neq x \in A_\mathfrak{p}$, then there is an s in $A - \mathfrak{p}$ such that $sx \in A$. Hence the set described contains $A_\mathfrak{p}$. The other inclusion holds as well. $\square$

Lemma 3.9. *Suppose $A_\mathfrak{p}$ is not a valuation ring. Then there is an x in K^* such that $(Ax \cap A) + (Ax^{-1} \cap A) \subseteq \mathfrak{p}$ and hence such that $A : \mathfrak{p} \subseteq (Ax \cap A) : (Ax \cap A)$.*

Proof. Since $A_\mathfrak{p}$ is not a valuation ring, there is an x in K^* such that neither x nor x^{-1} is in $A_\mathfrak{p}$. Hence $(Ax \cap A) + (Ax^{-1} \cap A) \subseteq \mathfrak{p}$ by Lemma 3.8. So $A : \mathfrak{p} \subseteq (A : (Ax \cap A)) \cap (A : (Ax^{-1} \cap A))$. Now $x(Ax^{-1} \cap A) = Ax \cap A$. This leads to $(A : (Ax \cap A)) \cap (A : (Ax^{-1} \cap A)) = (Ax \cap A) : (Ax \cap A)$. $\square$

Theorem 3.10 (Beck). *If $\mathfrak{p}$ is a prime ideal of A such that $A : \mathfrak{p} \not\subset \tilde{A}$, then $A_\mathfrak{p}$ is a principal valuation ring.* (Recall that $\tilde{A}$ is the complete integral closure of A.)

Proof. The above lemma shows that $A_\mathfrak{p}$ is a valuation ring (for if not, then $A : \mathfrak{p} \subseteq \mathfrak{a} : \mathfrak{a} \subseteq \tilde{A}$ for some fractionary ideal $\mathfrak{a}$).

$A_\mathfrak{p}$ is of rank one (i.e., $\mathfrak{p}A_\mathfrak{p}$ is a minimal prime ideal different from zero). For if not, then by Lemma 3.7, there is a prime ideal $\mathfrak{q} \subsetneqq \mathfrak{p}$ such that $A : \mathfrak{p} \subseteq \mathfrak{q} : \mathfrak{q} \subseteq \tilde{A}$, a contradiction.

Suppose $x \in (A : \mathfrak{p}) - \tilde{A}$. Then $x \notin \mathfrak{p} : \mathfrak{p}$. Since $x\mathfrak{p} \subseteq A$, it follows that $x\mathfrak{p}A_\mathfrak{p} \subseteq A_\mathfrak{p}$. If $x\mathfrak{p}A_\mathfrak{p} \subseteq \mathfrak{p}A_\mathfrak{p}$, then $x\mathfrak{p} \subseteq \mathfrak{p}A_\mathfrak{p} \cap A = \mathfrak{p}$, a contradiction. Hence $x\mathfrak{p}A_\mathfrak{p} = A_\mathfrak{p}$, so $\mathfrak{p}A_\mathfrak{p} = x^{-1}A_\mathfrak{p}$. Hence $A_\mathfrak{p}$ is a rank one valuation ring whose maximal ideal is principal. Thus $A_\mathfrak{p}$ is a principal valuation ring. $\square$

Corollary 3.11. *If A is completely integrally closed and $\mathfrak{p}$ is a prime ideal in A such that $A : \mathfrak{p} \not\subseteq \tilde{A}$, then $A_\mathfrak{p}$ is a principal valuation ring.* $\square$

Theorem 3.12. *Suppose A is completely integrally closed. If the set of divisorial ideals of A satisfies the maximum condition then the following hold:*

(a) *$\mathfrak{p} \in X^{(1)}(A)$ if and only $\mathfrak{p}$ is a maximal divisorial prime ideal.*

(b) *If $\mathfrak{p} \in X^{(1)}$, then $A_\mathfrak{p}$ is a principal valuation ring.*

(c) *$A = \bigcap\limits_{\mathfrak{p} \in X^{(1)}} A_\mathfrak{p}$.*

(d) *An ideal $\mathfrak{a}$ is divisorial if and only if $\mathfrak{a} = A : (A : (\mathfrak{p}_1 \ldots \mathfrak{p}_r))$ for some $\mathfrak{p}_i$ in $X^{(1)}$ and, when this happens, this "factorization" is unique.*

(e) *A divisorial ideal is contained in only finitely many $\mathfrak{p}$ in $X^{(1)}$.*

Proof. Beck has provided part of the proof. Let $\mathfrak{p}$ be a minimal nonzero prime ideal. Then $\mathfrak{p}$ is a maximal divisorial ideal. For suppose $f \in \mathfrak{p} - \{0\}$. Then $A : f^{-1}A \subseteq \mathfrak{p}$. Hence the family of divisorial ideals of the form $Ax^{-1} \cap A$ which are contained in $\mathfrak{p}$ is not empty. By the maximum condition, pick a maximal such divisorial ideal. Call this ideal $\mathfrak{a}$. We shall show it is a prime ideal.

For suppose f, g in A are such that $fg \in \mathfrak{a}$. Now $A(xf)^{-1} \cap A \supseteq \mathfrak{a}$, so $(A(xf)^{-1} \cap A) \subseteq \mathfrak{p}$ implies $\mathfrak{a} = A(xf)^{-1} \cap A$ by the maximality of $\mathfrak{a}$. Now $g \in A(xf)^{-1} \cap A$. So $A(xf)^{-1} \cap A$ contained in $\mathfrak{p}$ implies that $g \in \mathfrak{a}$. Suppose then that $A(xf)^{-1} \cap A \nsubseteq \mathfrak{p}$. Then there is an h in $A - \mathfrak{p}$ such that $hf \in A x^{-1} \cap A$. But $h(A(xh)^{-1} \cap A) \subseteq A x^{-1} \cap A \subseteq \mathfrak{p}$ and $h \notin \mathfrak{p}$. Therefore $A(xh)^{-1} \cap A \subseteq \mathfrak{p}$ and in particular $A(xh)^{-1} \cap A = \mathfrak{a}$. But then $f \in \mathfrak{a}$. Hence, in any case, one of f, g is in $\mathfrak{a}$. So $\mathfrak{a}$ is a prime ideal.

It has been seen that maximal divisorial ideals are prime ideals. Hence $X^{(1)}$ consists exactly of the maximal divisorial ideals and thus (a) is established.

Since A is completely integrally closed, $A_\mathfrak{p}$ is a principal valuation ring for each $\mathfrak{p}$ in $X^{(1)}$ by Beck's Theorem 3.10. This proves (b).

Now suppose $x \in \bigcap_{\mathfrak{p} \in X^{(1)}} A_\mathfrak{p}$. It is desired to prove that x is in A. If $\mathfrak{p} \in X^{(1)}$, then $A x^{-1} \cap A \not\subseteq \mathfrak{p}$ by Lemma 3.8. Hence $A x^{-1} \cap A$ is a divisorial ideal not contained in a maximal divisorial ideal. Thus $A x^{-1} \cap A = A$. In particular $x \in A$. So (c) is true.

To prove (d) suppose, by the maximum condition, that $\mathfrak{a}$ is a divisorial ideal which is not of the form $A:(A:(\mathfrak{p}_1 \ldots \mathfrak{p}_r))$ for suitable maximal divisorial ideals $\mathfrak{p}_1, \ldots, \mathfrak{p}_r$ and is maximal among such ideals. In particular, $\mathfrak{a}$ is not a maximal divisorial ideal. Hence there is a maximal divisorial ideal $\mathfrak{p}_1$ such that $\mathfrak{a} \subsetneqq \mathfrak{p}_1$. Hence $\mathfrak{a} \subsetneqq (A:\mathfrak{p}_1)\mathfrak{a} \subseteq (A:\mathfrak{p}_1)\mathfrak{p}_1 \subseteq A$. It cannot happen that equality is attained in the first position since $A:\mathfrak{p}_1 \subsetneqq A$. So $\mathfrak{a} \subsetneqq A:(A:(A:\mathfrak{p}_1)\mathfrak{a})$. Then, by the maximality of $\mathfrak{a}$, this second ideal is of the desired form, say equal to $A:(A:(\mathfrak{p}_2 \ldots \mathfrak{p}_r))$. By several applications of the equality $\mathfrak{a}:\mathfrak{b}\mathfrak{c} = (\mathfrak{a}:\mathfrak{b}):\mathfrak{c}$, it follows that $\mathfrak{a} = A:(A:(\mathfrak{p}_1 \ldots \mathfrak{p}_r))$ and thus $\mathfrak{a}$ is indeed of the desired form.

To see the uniqueness, note that $\mathfrak{p}_1 \ldots \mathfrak{p}_r \subseteq \mathfrak{a}$. If $\mathfrak{a} \subseteq \mathfrak{p} \in X^{(1)}$, then $\mathfrak{p}$ is one of the $\mathfrak{p}_i$. Hence $\mathfrak{a}$ is contained in at most a finite number of prime divisorial ideals (giving (e)) and if $\mathfrak{a} \subseteq \mathfrak{p}_1$ then $\mathfrak{a}:\mathfrak{p}_1 = A:(A:(\mathfrak{p}_2 \ldots \mathfrak{p}_r))$ showing the uniqueness. So both (d) and (e) have been verified. $\square$

Corollary 3.13. *If A is completely integrally closed and its fractionary ideals have the maximum condition, then A is a Krull domain.*

Corollary 3.14. *Under the same hypotheses, $D(A)$ is free on the prime ideals in $X^{(1)}$.*

These follow from (b) and (c) and (d) respectively. $\square$

Let A be a Krull domain. If Y is a subset of $X^{(1)}(A)$, then the *subintersection B*, defined to be $\bigcap_{\mathfrak{p} \in Y} A_\mathfrak{p}$, is again a Krull domain according to Corollary 1.5. If $\mathfrak{P} \in X^{(1)}(B)$, then $B_\mathfrak{P}$ is a principal valuation ring which is a subintersection $\bigcap_{\mathfrak{p} \in Y'} A_\mathfrak{p}$ where $Y' = \{\mathfrak{p} \in Y : B - \mathfrak{P} \subseteq A_\mathfrak{p}^*\}$.

Hence there is at most one $\mathfrak{p}$ in Y such that $B_{\mathfrak{P}} = A_{\mathfrak{p}}$. Thus we can state the next result.

Proposition 3.15. *Suppose* $Y \subseteq X^{(1)}(A)$. *Let* B *be the subintersection* $\bigcap\limits_{\mathfrak{p} \in Y} A_{\mathfrak{p}}$. *Then the homomorphism* $\operatorname{Spec} B \to \operatorname{Spec} A$ *induces a bijection* $X^{(1)}(B) \to Y$. $\square$

§ 4. Krull's Normality Criterion and the Mori-Nagata Integral Closure Theorem

Let A be a noetherian integral domain. Then each fractional ideal is of finite type. So an element in K^* is almost integral over A if and only if it is integral over A. Thus A is normal if and only if it is completely integrally closed. The ideals of A, and in particular, the divisorial ideals satisfy the ascending chain condition. So Theorem 3.12 implies that a noetherian normal integral domain is a Krull domain. This is part of Krull's normality theorem [Krull, 1931]. (Serre has generalized this result to noetherian rings which are not necessarily integral domains. See [EGA, 1965]).

Theorem 4.1. *Let* A *be a noetherian integral domain. Then* A *is normal if and only if* A *satisfies the conditions:*

(S_2) depth $A_{\mathfrak{q}} \geqslant \inf(2, \operatorname{ht} \mathfrak{q})$ *for all* $\mathfrak{q}$ *in* $\operatorname{Spec} A$.

(R_1) *If* $\mathfrak{p} \in X^{(1)}(A)$, *then* $A_{\mathfrak{p}}$ *is a regular local ring* (i.e., a principal valuation ring).

Proof. If A is normal, then (R_1) is a consequence of Theorem 3.12. Suppose that $\mathfrak{q}$ is a prime ideal with $\operatorname{ht} \mathfrak{q} \geqslant 2$. Then $A_{\mathfrak{q}}$ is normal. To verify (S_2) (which is true if $\operatorname{ht} \mathfrak{q} < 2$ since then $A_{\mathfrak{q}}$ is regular), a regular $A_{\mathfrak{q}}$-sequence of length 2 must be found in $\mathfrak{q} A_{\mathfrak{q}}$. Let f be in $\mathfrak{q} A_{\mathfrak{q}}$ with $f \neq 0$. Then there are at most a finite number of elements in $X^{(1)}(A_{\mathfrak{q}})$ which contain f. It cannot happen that these finite number of prime ideals of height one exhaust $\mathfrak{q}$ since $\operatorname{ht} \mathfrak{q} > 1$. Let g in $\mathfrak{q} A_{\mathfrak{q}}$ be an element not in any of these prime ideals of height one which contain f. Then f, g is a regular $\mathfrak{q} A_{\mathfrak{q}}$-sequence. For let a, b in $A_{\mathfrak{q}}$ be such that $af = bg$. Then $af A_{\mathfrak{p}} = bg A_{\mathfrak{p}}$ for all prime ideals $\mathfrak{p}$ in $X^{(1)}(A_{\mathfrak{q}})$. If $f \in \mathfrak{p}$, then $g A_{\mathfrak{p}} = A_{\mathfrak{p}}$ so $a/g = b/f$ in $A_{\mathfrak{p}}$. If $f \notin \mathfrak{p}$, then f is a unit in $A_{\mathfrak{p}}$ so $b/f \in A_{\mathfrak{p}}$. Hence $(b/f) \in \bigcap A_{\mathfrak{p}} = A_{\mathfrak{q}}$. Thus $b \in f A_{\mathfrak{q}}$. This shows that f, g is a regular $\mathfrak{q} A_{\mathfrak{q}}$-sequence.

Now suppose A is a noetherian domain which satisfies (S_2). Using some elementary properties of associated prime ideals we show that (S_2) implies $A = \bigcap\limits_{\mathfrak{p} \in X^{(1)}} A_{\mathfrak{p}}$. For suppose x is in the intersection. Write $x = f/g$ with f, g in A and $fg \neq 0$. Let $\mathfrak{a} = A x^{-1} \cap A$ which is

$\{a \in A : af \in g A\}$. Let q be a prime ideal containing $\mathfrak{a}$ minimally. It cannot happen that ht q < 2. For $\mathfrak{a}_q = \{r \in A_q : rx \in A_q\}$. If ht q < 2, then A_q is one of the rings containing x and $\mathfrak{a}_q = A_q$. Now depth $A_q \geqslant 2$ otherwise, so, since $\mathfrak{a}_q$ is a primary $q A_q$-ideal, depth$_{\mathfrak{a}_q} A_q \geqslant 2$, and $\mathfrak{a}_q$ contains a regular A_q-sequence of length 2, say u, v. Now $u f = g r$ and $v f = g s$ with r, s in A_q. Hence $us = vr$. But this implies that $r \in u A_q$ and consequently $f \in g A_q$. Hence $\mathfrak{a}_q = A_q$. So no prime ideal contains $\mathfrak{a}$ and thus $\mathfrak{a} = A$ which implies $x \in A$.

If (R_1) holds as well, then A is the intersection of principal valuation rings, so A is normal. □

Corollary 4.2 (To the proof). *Let A be a noetherian integral domain satisfying* (S_2). *Let M be an A-lattice in V. Then* $A : (A : M) = \bigcap_{\mathfrak{p} \in X^{(1)}} A : (A : M)_{\mathfrak{p}}$.
If $A_{\mathfrak{p}}$ is Gorenstein for each $\mathfrak{p}$ in $X^{(1)}$, then $A : (A : M) = \bigcap M_{\mathfrak{p}}$.

Proof. We have $A : M \subseteq \bigcap (A : M)_{\mathfrak{p}}$ which is just $\bigcap (A_{\mathfrak{p}} : M_{\mathfrak{p}})$. Suppose $\alpha \in \bigcap A_{\mathfrak{p}} : M_{\mathfrak{p}}$. Let m be in M. Then $M \subseteq M_{\mathfrak{p}}$, so $\alpha(m) \in A_{\mathfrak{p}}$ for each $\mathfrak{p} \in X^{(1)}$. Hence $\alpha(m) \in \bigcap A_{\mathfrak{p}} = A$. So $A : M = \bigcap A_{\mathfrak{p}} : M_{\mathfrak{p}}$ and hence $A : (A : M) = \bigcap A_{\mathfrak{p}} : (A_{\mathfrak{p}} : M_{\mathfrak{p}})$.
If $A_{\mathfrak{p}}$ is Gorenstein, then $M_{\mathfrak{p}} = A_{\mathfrak{p}} : (A_{\mathfrak{p}} : M_{\mathfrak{p}})$. □

The next question which arises is: Is the integral closure of a Noetherian integral domain noetherian? The answers are as follows: If the Krull dimension is one, then every ring between A and its integral closure is noetherian (Krull-Akizuki Theorem). If the Krull dimension is two, then the integral closure is noetherian [Nagata, 1962] but there can be a non-noetherian ring between A and its integral closure. If the Krull dimension is three, then the integral closure need not be noetherian [Nagata, loc. cit.]. Krull conjectured that in any case the integral closure of a noetherian integral domain is a finite discrete principal order (i.e., a Krull domain).

Mori [1952—53] affirmed the conjecture for local rings. Then Nagata [1955] descended Mori's result from the local case to the general case.

Theorem 4.3 (Mori-Nagata integral closure theorem). *Suppose A is the integral closure of the noetherian integral domain B. Then*
(a) *A is a Krull domain.*
(b) *Spec $A \xrightarrow{i}$ Spec B has finite fibres (i.e., $i^{-1}(\mathfrak{p})$ is a finite set for each $\mathfrak{p}$ in Spec B).*
(c) *If $\mathfrak{P} \in$ Spec A and $\mathfrak{p} = \mathfrak{P} \cap B$, then $[A_{\mathfrak{p}}/\mathfrak{P} A_{\mathfrak{p}} : B_{\mathfrak{p}}/\mathfrak{p} B_{\mathfrak{p}}]$ is finite* (i.e., the stalks $k(\mathfrak{P})$ are finite field extensions of $k(\mathfrak{p})$).

The proof is divided into two parts.

Theorem 4.4 (Mori). *If B is a local noetherian integral domain, then* (a), (b), (c) *above hold.*

Proof (Nagata). Let $\hat{B}$ denote the completion of B in its $\mathfrak{m}(B)$-adic topology. Let $\mathfrak{n}$ denote the nilradical of B and set $B' = \hat{B}/\mathfrak{n}$.

Since B is an integral domain, the contraction $\mathfrak{n} \cap B = (0)$, so B is embedded in B'. The reduction B' is reduced (has zero nilradical), so its total ring of quotients, say L, is a product of fields, $L = \prod_{i=1}^{n} L_i$, where $L_i = B'_{\mathfrak{p}_i}$ for the minimal prime ideal $\mathfrak{p}_i$ of B'. Also K is a K-subalgebra of each L_i and L is the product of the L_i as K-algebras.

Let I_i denote the integral closure of $B_i = B'/\mathfrak{p}_i$ in L_i for each i. Then the integral closure I of B' in L is just $\prod_{i=1}^{n} I_i$. The first thing to note about I is that $I \cap K = A$, the integral closure of B in K. The containment $A \subseteq I \cap K$ is clear. Suppose, on the other hand, that x in K is integral over B', say $x^m + b'_{m-1} x^{m-1} + \cdots + b'_0 = 0$, with each b'_j in B'. Let b_j in $\hat{B}$ be a preimage of b'_j in B' for $j = 0, \ldots, m-1$. Also write $x = f/g$ with f, g in B, $fg \neq 0$.

$$\begin{array}{ccccccc}
\hat{B} & \longrightarrow & B' & \hookrightarrow & \prod B'/\mathfrak{p}_i & \hookrightarrow & I = \prod I_i & \hookrightarrow & L = \prod L_i \\
\uparrow & & & & \uparrow & & \uparrow \\
B & \hookrightarrow & & & A = I \cap K & \hookrightarrow & K
\end{array}$$

Fig. 1

Then $f^m + b_{m-1} f^{m-1} g + \cdots + b_1 f g^{m-1} + b_0 g^m$ is in the nilradical $\mathfrak{n}$, so it is nilpotent. Hence there is a relation

$$f^s + d_{s-1} f^{s-1} g + \cdots + d_1 f g^{s-1} + d_0 g^s = 0$$

with d_j in $\hat{B}$ for each j. So f^s is in the ideal $\sum_{i=1}^{s} \hat{B} f^{s-i} g^i$. Since $\hat{B}$ is a faithfully flat B-module and f and g are in B, it follows that $\left(\sum_{1}^{s} \hat{B} f^{s-i} g^i \right) \cap B = \sum_{1}^{s} B f^{s-i} g^i$. Hence $f/g = x$ is integral over B.

Recall now the crucial result [Nagata, 1962, 32.1, page 112].

Let R be a complete local integral domain. Suppose S is an integral domain, integral over R, whose field of quotients is a finite extension of the field of quotients of R. Then S is an R-module of finite type.

An application of this result yields that each I_j is a noetherian integral domain which is normal, hence a Krull domain. By the lying over theorem, there are only finitely many prime ideals of I_j which intersect to any given prime ideal of $B'/\mathfrak{p}_j$. Also by the integrality and finiteness of I_j over $B'/\mathfrak{p}_j$, the residue fields must be finite extensions.

Now $A = K \cap I = \bigcap_{j=1}^{n} K \cap I_j$. But $K \cap I_j$ is a Krull domain by Proposition 1.2, so A is a Krull domain, it being a finite intersection of Krull domains.

It remains to show that A has finite number of maximal ideals.

Let $\mathfrak{m}_1, \ldots, \mathfrak{m}_m$ be distinct maximal ideals in A. For each i, pick t_i in $\mathfrak{m}_i - \left(\bigcup_{i \neq j} \mathfrak{m}_j \right)$. Consider the noetherian integral domain $B[t_1, \ldots, t_m]$, which has a finite number of maximal ideals (in fact exactly m of them). Its completion with respect to its Jacobson radical is just $\hat{B}[[t_1, \ldots, t_m]]$, which is a product of m complete local rings. But there can be no more idempotents in $\hat{B}[[t_1, \ldots, t_m]]/\mathfrak{n}'$ ($\mathfrak{n}'$ being the nilradical), than there are in I. Hence m is finite.

Now the various parts of the theorem are verified for the local ring B and the maximal ideals of A. To complete the proof in the local case, suppose $\mathfrak{P}$ is a prime ideal of A. Let $\mathfrak{p} = \mathfrak{P} \cap A$. Then the integral closure of the local ring $B_\mathfrak{p}$ is $A_\mathfrak{p}$ which, by the theorem applied to the local ring $B_\mathfrak{p}$, has only a finite number of prime ideals lying over $\mathfrak{p} B_\mathfrak{p}$. And each of them satisfies the finiteness condition (c). Thus Theorem 4.4 is established. □

To go from the local to the global version, a lemma is needed.

Lemma 4.5 (Beck). *Let A be the integral closure of the noetherian integral domain B. Let $\{a_1, \ldots, a_r\}$ be a set of elements in A. Then $B:(B:\sum Ba_j) \subseteq A:(A:\sum A a_j)$.*

Proof. Since A is the integral closure of B, there is a fractionary B-ideal $\mathfrak{a}$ such that $a_i \mathfrak{a} \subseteq \mathfrak{a}$ for each i. Hence $a_i(B:\mathfrak{a}) \subseteq B:\mathfrak{a}$ and so $a_i(B:(B:\mathfrak{a})) \subseteq B:(B:\mathfrak{a})$. Hence $\mathfrak{a}$ may be assumed to be divisorial.

The first step is to show that $B:(B:\sum Ba_i) \subseteq A$. Let y be in $B:(B:\sum Ba_i)$. Then $B y^{-1} = B:By \supseteq B:\sum Ba_i = \bigcap_i Ba_i^{-1}$. Hence $By^{-1} \cap B \supseteq B \cap \left(\bigcap_i Ba_i^{-1} \right)$. As $a_i \mathfrak{a} \subseteq \mathfrak{a}$, it follows that $\mathfrak{a} = (a_i^{-1} \mathfrak{a}) \cap \mathfrak{a} = (Ba_i^{-1} \cap B):(B:\mathfrak{a})$ for each i. Hence $\mathfrak{a} = (Ba_i^{-1} \cap \cdots \cap Ba_r^{-1} \cap B):(B:\mathfrak{a}) \subseteq (By^{-1} \cap B):(B:\mathfrak{a})$. But $(By^{-1} \cap B):(B:\mathfrak{a}) = \mathfrak{a} y^{-1} \cap \mathfrak{a} \subseteq \mathfrak{a}$. These containments imply: $\mathfrak{a} = \mathfrak{a} y^{-1} \cap \mathfrak{a}$ or $y\mathfrak{a} \subseteq \mathfrak{a}$ or yet further $y \mathfrak{a} \subseteq \mathfrak{a}$. Hence y is integral over B, which implies that $y \in A$.

The second step, then, is to show that $(B:(B:\sum Ba_i))(A:\sum Aa_i)\subseteq A$. Suppose $z(\sum Aa_i)\subseteq A$ (i.e., $z\in A:(\sum Aa_i)$). Then $za_i\in A$ for each i. Hence $z(B:(B:\sum Ba_i))=B:(B:\sum Ba_iz)$. By the first step (with a_i replaced by a_iz) $B:(B:\sum Ba_iz)\subseteq A$, which is the desired result. $\square$

Proposition 4.6. *Let A be the integral closure of the noetherian integral domain B. If $\mathfrak{P}$ is a divisorial* (prime) *ideal in A then $\mathfrak{P}\cap B$ is a divisorial* (prime) *ideal in B.*

Proof. Let $\mathfrak{p}=\mathfrak{P}\cap B$. Then $\mathfrak{p}$ is of finite type as a B-module. Hence $B:(B:\mathfrak{p})\subseteq A:(A:A\mathfrak{p})$ by the lemma. But $\mathfrak{p}A\subseteq\mathfrak{P}$ and so $A:(A:A\mathfrak{p})$ $\subseteq A:(A:\mathfrak{P})=\mathfrak{P}$. Hence $B:(B:B\mathfrak{p})\subseteq\mathfrak{P}$. But also $B:(B:\mathfrak{p})\subseteq B$, so $B:(B:\mathfrak{p})$ $\subseteq\mathfrak{P}\cap B=\mathfrak{p}$. Therefore the ideal $\mathfrak{p}=B:(B:\mathfrak{p})$, which establishes the proposition. $\square$

Return now to the proof of Theorem 4.3. Let A be the integral closure of the noetherian integral domain B. It must be shown that A is a Krull domain. The other two statements of the theorem are local in nature and follow from the local case, Theorem 4.4.

Let $\mathfrak{m}$ be a maximal ideal of B. Then $A_\mathfrak{m}$ is the integral closure of $B_\mathfrak{m}$, which, by Mori's theorem, is a Krull domain. As $A=\bigcap_\mathfrak{m} A_\mathfrak{m}$, the ring A is the intersection of principal valuation rings. Suppose $\mathfrak{P}\in X^{(1)}(A_\mathfrak{p})$ with $\mathfrak{p}=\mathfrak{P}\cap A$. Then $\mathfrak{P}A_\mathfrak{p}\in X^{(1)}(A_\mathfrak{p})$, which implies that $A_\mathfrak{P}$ is a principal valuation ring. Since each $\mathfrak{P}$ in $X^{(1)}(A)$ is contained in a maximal ideal, since each element of $X^{(1)}(A_\mathfrak{M})$ restricts to an element of $X^{(1)}(A)$ (where $\mathfrak{M}$ is a prime ideal in A) and since $A_\mathfrak{M}$ is a Krull domain, it follows that $A=\bigcap_{\mathfrak{P}\in X^{(1)}} A_\mathfrak{P}$.

It remains to verify the finite character property for this family $\{A_\mathfrak{P}:\mathfrak{P}\in X^{(1)}(A)\}$ of principal valuation rings. For this, suppose $f\in A$ with $f\neq 0$. Since f is integral over B, the finite extension, $B[f]$ is a noetherian ring whose integral closure is A. If $\mathfrak{P}\in X^{(1)}(A)$, then $\mathfrak{P}=\bigcap_{\mathfrak{Q}\in X^{(1)}} \mathfrak{P}A_\mathfrak{Q}$. Hence $\mathfrak{P}$ is a divisorial prime ideal in A. By Proposition 4.6 $\mathfrak{P}\cap B[f]$ is a divisorial prime ideal of $B[f]$. Furthermore, if $f\in\mathfrak{P}$, then $f\in\mathfrak{P}\cap B[f]$. Denote $\mathfrak{P}\cap B[f]$ by $\mathfrak{p}$.

Since $B[f]$ is noetherian, we get $B[f]:\mathfrak{p}=B[f]+\sum_1^m x_iB[f]$ with x_i in K. Then $\mathfrak{p}=B[f]:(B[f]:\mathfrak{p})=B[f]:(B[f]+\sum_1^m x_iB[f])$ $=\bigcap_1^m(B[f]\cap x_i^{-1}B[f])$. As $\mathfrak{p}$ is a prime ideal, it contains one of the terms in this intersection. Hence $\mathfrak{p}=B[f]\cap x^{-1}B[f]$ for some x in K. Write $x=b/c$ with b and c in $B[f]$. Then $\mathfrak{p}=\{g\in B[f]:gb\in cB[f]\}$. Since $f\in\mathfrak{p}$, it follows that $fb=ch$ for some h in $B[f]$. Hence $\mathfrak{p}=\{g\in B[f]:gh\in fB[f]\}$. Thus $\mathfrak{p}\in\text{Ass}(B[f]/fB[f])$.

The conclusion one draws from the above paragraph is this: If $\mathfrak{P}$ in $X^{(1)}(A)$ contains f then $\mathfrak{P} \cap B[f] \in \mathrm{Ass}(B[f]/fB[f])$. As there are at most a finite number of prime ideals in $\mathrm{Ass}(B[f]/fB[f])$ ($B[f]$ is noetherian), and as there are at most a finite number of prime ideals of A lying over any prime ideal of $B[f]$, we see that $\{\mathfrak{P} \in X^{(1)}(A): f \in \mathfrak{P}\}$ is a finite set. Thus the finite character property is established. $\square$

It may be of some interest to mention Akizuki's [1958] conclusion of the proof of Theorem 4.3, namely the verification of the finite character property.

Let $\mathfrak{P}$ be in $X^{(1)}(A_{\mathfrak{m}})$ where $\mathfrak{m}$ is a maximal ideal of B. Let $\mathfrak{p} = \mathfrak{P} \cap B$. Suppose $\mathfrak{Q}$ is a nonzero prime ideal of $A_{\mathfrak{n}}$ where $\mathfrak{n}$ is some other maximal ideal of B. If $\mathfrak{Q} \cap B = \mathfrak{p}$, then $\mathfrak{Q} A_{\mathfrak{p}} \cap B_{\mathfrak{p}} = \mathfrak{p} B_{\mathfrak{p}} = \mathfrak{P} A_{\mathfrak{p}} \cap B_{\mathfrak{p}}$. As $A_{\mathfrak{p}}$ is the integral closure of $B_{\mathfrak{p}}$, the ideals $\mathfrak{P} A_{\mathfrak{p}}$ and $\mathfrak{Q} A_{\mathfrak{p}}$ are maximal ideals of $A_{\mathfrak{p}}$ (by the Krull-Cohen-Seidenberg Theorem). By Theorem 4.4, there are but a finite number of minimal nonzero prime ideals of A which lie over a given prime ideal in B.

Suppose $f \in A$, $f \neq 0$. Let f be in $\mathfrak{P}$ where $\mathfrak{P} \in X^{(1)}(A)$. The ideal $\mathfrak{q}(\mathfrak{P}) = \{g \in B[f]: g A_{\mathfrak{P}} \in f A_{\mathfrak{P}}\}$ is a primary ideal with radical $\mathfrak{p} = \mathfrak{P} \cap B[f]$. Then it follows that $fA \cap B[f] = \bigcap_{f \in \mathfrak{P}} \mathfrak{q}(\mathfrak{P})$. This ideal has only finitely many ideals in its primary decomposition. Thus there is but a finite number of $\mathfrak{P}$ with f in $\mathfrak{P}$.

This shows the finite character property of A. It avoids the use of Beck's proposition as well as the use of completions.

§ 5. Divisorial Lattices and the Approximation Theorem

We have seen that a Krull domain A is the intersection of its localizations at its minimal nonzero prime ideals, that each of these is a principal valuation ring and that each nonzero element is in only a finite number of these prime ideals. We next consider characterizations of divisorial A-lattices which take the same form. We will show that an A-lattice M is divisorial if and only if $M = \bigcap_{\mathfrak{p} \in X^{(1)}} M_{\mathfrak{p}}$. As a corollary to this, we obtain the result: $S^{-1}(N:M) = (S^{-1}N):(S^{-1}M)$ for a divisorial lattice N. Although not included here, these results have applications to the theory of orders over Krull domains. The interested reader is referred to [Fossum, 1968].

Before going into the properties which (divisorial) lattices enjoy, a result concerning successive localizations is given.

Lemma 5.1. *Let A be an integral domain. Suppose $\mathfrak{p}$ and $\mathfrak{q}$ are prime ideals in A such that $\mathfrak{p} \not\subseteq \mathfrak{q}$. If $A_\mathfrak{p}$ is a principal valuation ring, then $(A_\mathfrak{p})_\mathfrak{q} = K$. If M is an $A_\mathfrak{p}$ module where $A_\mathfrak{p}$ is a principal valuation ring, then $M_\mathfrak{q} = K \otimes_{A_\mathfrak{p}} M$.*

Proof. The second statement follows from the first. Since $A_\mathfrak{p}$ is a principal valuation ring, the field $K = (A_\mathfrak{p})_f$ for any nonunit f in $A_\mathfrak{p}$. If $f \in \mathfrak{p} - \mathfrak{q}$, then $(A_\mathfrak{p})_f \subseteq (A_\mathfrak{p})_\mathfrak{q} \subseteq K$. ☐

Most of the results here are straightforward generalizations of the corresponding results for lattices over noetherian Krull domains which are found in [Bourbaki, 1965]. For a Krull domain which is not necessarily noetherian, the notion of associated prime ideals and the corresponding decomposition theory is found in [Claborn and Fossum, 1968].

Proposition 5.2. *Let A be a Krull domain with field of quotients K. Let $Z = X^{(1)}(A)$. Let V be a finite dimensional vector space over K and M an A-lattice in V.*

(a) If N is an A-lattice in V, then $N_\mathfrak{p} = M_\mathfrak{p}$ for almost all $\mathfrak{p}$ in Z.

(b) If, for each $\mathfrak{p}$ in Z, there is given an $A_\mathfrak{p}$-lattice (necessarily free) $N(\mathfrak{p})$ such that $N(\mathfrak{p}) = M_\mathfrak{p}$ for almost all $\mathfrak{p}$, then $N = \bigcap_\mathfrak{p} N(\mathfrak{p})$ is an A-lattice in V and is the maximal A-lattice in V such that $N_\mathfrak{p} = N(\mathfrak{p})$ for all $\mathfrak{p}$ in Z.

(c) M is a divisorial A-lattice in V if and only if $M = \bigcap_{\mathfrak{p} \in Z} M_\mathfrak{p}$.

Proof. (a) Since M and N are lattices in V, there are nonzero f and g in A such that $fM \subseteq N \subseteq g^{-1}M$. Since fg is a unit in $A_\mathfrak{p}$ for almost all $\mathfrak{p}$ in Z, it is the case that $M_\mathfrak{p} = N_\mathfrak{p}$ for almost all $\mathfrak{p}$.

(b) If L is any free A-lattice, then $L_\mathfrak{p} = M_\mathfrak{p}$ for almost all $\mathfrak{p}$. By multiplying by a suitable element in K^*, it may be assumed that $N(\mathfrak{p}) \subseteq L_\mathfrak{p}$ for all $\mathfrak{p}$ in Z with equality for almost all $\mathfrak{p}$.

Now $L = \bigcap_\mathfrak{p} L_\mathfrak{p}$ since L is free. To see this, let $v_1, \ldots, v_n$ be a basis for L. Suppose $w \in \bigcap_\mathfrak{p} L_\mathfrak{p}$. Write $w = x_1 v_1 + \cdots + x_n v_n$ with x_i in K for each i. Since $w \in L_\mathfrak{p}$, each $x_i \in A_\mathfrak{p}$ for all $\mathfrak{p}$. Hence each $x_i \in \bigcap_\mathfrak{p} A_\mathfrak{p} = A$. So $w \in L$.

Since each $N(\mathfrak{p}) \subseteq L_\mathfrak{p}$, it follows that $N = \bigcap_\mathfrak{p} N(\mathfrak{p}) \subseteq L$. The same argument (for a different element in K^*) can be used to show that $N \supseteq L'$ for some free A-lattice L'. Hence N is an A-lattice in V.

Now suppose N' is an A-lattice in V such that $N'_\mathfrak{p} = N(\mathfrak{p})$ for each $\mathfrak{p}$ in Z. Then $N' \subseteq \bigcap_\mathfrak{p} N'_\mathfrak{p} = \bigcap_\mathfrak{p} N(\mathfrak{p}) = N$. So N is the maximal such lattice provided $N_{\mathfrak{p}} = N(\mathfrak{p})$ for all $\mathfrak{p}$. But it is to establish this equality that we stated Lemma 5.1. Now $N = N(\mathfrak{p}_1) \cap N(\mathfrak{p}_2) \cap \cdots \cap N(\mathfrak{p}_r) \cap L$ where $\mathfrak{p}_1, \ldots, \mathfrak{p}_r$ are exactly those prime ideals in Z for which $N(\mathfrak{p}) \neq L_{\mathfrak{p}}$. Furthermore $N_{\mathfrak{p}} = N(\mathfrak{p}_1)_{\mathfrak{p}} \cap \cdots \cap N(\mathfrak{p}_r)_{\mathfrak{p}} \cap L_{\mathfrak{p}}$. If $\mathfrak{p}$ is not one of the $\mathfrak{p}_j$, then $N(\mathfrak{p}_j)_{\mathfrak{p}} = V$ for all $\mathfrak{p}_j$ by Lemma 5.1. Hence $N_{\mathfrak{p}} = L_{\mathfrak{p}} = N(\mathfrak{p})$. Also $N(\mathfrak{p}_i)_{\mathfrak{p}_j} = V$ if $i \neq j$. Hence $N_{\mathfrak{p}_j} = N(\mathfrak{p}_j)_{\mathfrak{p}_j} \cap L_{\mathfrak{p}_j}$. As $N(\mathfrak{p}_j)$ is already an $A_{\mathfrak{p}_j}$-module, we get $N(\mathfrak{p}_j) = N(\mathfrak{p}_j)_{\mathfrak{p}_j}$. So $N_{\mathfrak{p}_j} = N(\mathfrak{p}_j)$.

(c) We continue the proof of (c) after the proof of the next lemma.

Lemma 5.3. *Let M be an A-lattice. Then $A:M = \bigcap_{\mathfrak{p}} A_{\mathfrak{p}}:M_{\mathfrak{p}}$.*

Proof. It is the case that $(A:M)_{\mathfrak{p}} \subseteq A_{\mathfrak{p}}:M_{\mathfrak{p}}$. For if $\alpha(M) \subseteq A$ and $s \notin \mathfrak{p}$, then $\alpha(M_{\mathfrak{p}}) \subseteq A_{\mathfrak{p}}$ and $s^{-1}\alpha(M_{\mathfrak{p}}) \subseteq A_{\mathfrak{p}}$. Hence $s^{-1}\alpha \in A_{\mathfrak{p}}:M_{\mathfrak{p}}$. So $A:M \subseteq \bigcap(A:M)_{\mathfrak{p}} \subseteq \bigcap A_{\mathfrak{p}}:M_{\mathfrak{p}}$. Now suppose $\alpha \in \bigcap A_{\mathfrak{p}}:M_{\mathfrak{p}}$ (which is a lattice in $\mathrm{Hom}_K(V,K)$ by (b)). Since $M \subseteq M_{\mathfrak{p}}$, it follows that $\alpha(M) \subseteq A_{\mathfrak{p}}$ for each p and hence that $\alpha(M) \subseteq \bigcap A_{\mathfrak{p}} = A$. □

Corollary 5.4. *For each A-lattice M, the localization $(A:M)_{\mathfrak{p}} = A_{\mathfrak{p}}:M_{\mathfrak{p}}$ for all $\mathfrak{p}$ in Z.*

Proof. This follows from part (b) of Proposition 5.2. □

Continuation of the proof of (c). Suppose $M = \bigcap M_{\mathfrak{p}}$. Then $A:(A:M) = \bigcap A_{\mathfrak{p}}:(A_{\mathfrak{p}}:M_{\mathfrak{p}})$ by Lemma 5.3 and its Corollary 5.4. But $A_{\mathfrak{p}}:(A_{\mathfrak{p}}:M_{\mathfrak{p}}) = M_{\mathfrak{p}}$. Hence $M = A:(A:M)$ so M is divisorial.

Conversely suppose M is divisorial. Then by Lemma 5.3, the lattice $A:M = \bigcap A_{\mathfrak{p}}:M_{\mathfrak{p}}$. Let $N = A:M$. Then $N = \bigcap N_{\mathfrak{p}}$. Hence by Lemma 5.3, we get the equalities $M = A:N = \bigcap A_{\mathfrak{p}}:N_{\mathfrak{p}} = \bigcap A_{\mathfrak{p}}:(A_{\mathfrak{p}}:M_{\mathfrak{p}}) = \bigcap M_{\mathfrak{p}}$. □

Corollary 5.5. *Let A be a Krull domain as in the theorem.*

(a) *The homomorphism $c_M : M \to \mathrm{Hom}_A(\mathrm{Hom}_A(M,A),A)$ is an isomorphism if and only if $M = \bigcap M_{\mathfrak{p}}$.*

(b) *A fractionary ideal $\mathfrak{a}$ is divisorial if and only if $\mathfrak{a} = \bigcap \mathfrak{a}_{\mathfrak{p}}$.*

(c) *Let S be a multiplicatively closed subset of A. Let N be a divisorial A-lattice and M an A-lattice. Then $S^{-1}(N:M) = S^{-1}N : S^{-1}M$.*

(d) *The divisorial A-lattices in an A-lattice M satisfy the ascending chain condition.*

(e) *The A-lattice M is divisorial if and only if $\mathrm{Ann}_A(Ax)$ is a divisorial ideal for each x in KM/M.*

(f) *The A-lattice M is divisorial if and only if every regular A-sequence of length two is a regular M-sequence.*

Proof. (a) and (b) are clear. (c) is proved in [Bass, 1968]. To see (d), suppose that $N_1 \subseteq \cdots \subseteq M$ is an ascending chain of divisorial A-lattices in KM. Then $\{(N_i)_p\}$ is an ascending chain of A_p-lattices in M_p. But M_p is a noetherian A_p-module. Furthermore, $(N_1)_p = M_p$ for almost all p in Z. Hence the chain stops ascending as in the proof of Theorem 3.6.

(e) Let M be an A-lattice in KM. Suppose $x \in \bigcap M_p$. Let $a(x) = \{f \in A : fx \in M\}$. Then $a(x)$ is the annihilator of the coset of x in KM/M. If it is assumed that each such annihilator is divisorial, then $a(x)$ is divisorial. Suppose that $a(x) \neq A$ for some such x (i.e., $x \notin M$). Pick x such that $a(x)$ is maximal. Let p in Z contain $a(x)$. Then $a(x)_p = \{f \in A_p : fx \in M_p\}$. (This follows since $a(x)$ is a maximal annihilator. All that really needs to be shown is that $a(x)_p$ contains this set. Suppose, then, that f in A and s in $A-p$ are such that $s^{-1}fx \in M_p$. Then $s^{-1}fx = t^{-1}y$ for some t in $A-p$ and y in M. Hence $ftx = sy \in M$. Thus $f \in a(tx)$ which contains $a(x)$. However $a(tx) = A$ would give tx in M or t in $a(x)$ which is contained in p, a contradiction. Thus $a(tx) = a(x)$, which gives f in $a(x)$ and consequently $s^{-1}f$ in $a(x)_p$.) But $x \in M_p$, so $a(x)_p = A_p$ which is a contradiction to $a(x) \subseteq p$. Hence each such $a(x)$ divisorial implies that $M = \bigcap M_p$.

Now suppose that M is divisorial. Let v be in KM and define $a(v)$ as above. We wish to show that $a(v)$ is divisorial. Suppose that $f \in \bigcap a(v)_p$. Then it follows that $fv \in M_p$ for each p. Hence $fv \in \bigcap M_p = M$. So $f \in a(v)$. This shows that $a(v)$ is divisorial.

(f) Suppose f, g is a regular A-sequence and M is divisorial. Suppose $fx + gy = 0$ where $x, y \in M$. Since M is divisorial, the lattice fM is divisorial. Let $\bar{y}$ denote the class of y in KM/fM. Then $\mathrm{Ann}_A(A\bar{y})$ is divisorial. But $\mathrm{Ann}_A(A\bar{y})$ contains both f (since $fy \in fM$) and g (since $gy = -fx \in fM$). The only divisorial ideal of A which contains a regular A-sequence of length two is A. Hence $\bar{y} = 0$, so $y \in fM$. Conversely, suppose every regular A-sequence of length two is a regular M-sequence. Suppose $x \in \bigcap M_p$. Then $a(x) = \{f \in A : fx \in M\}$. Since Ax is free, the ideal $a(x)_p = \{f \in A_p : fx \in M_p\}$ for all p in Z. But $a(x)_p = A_p$. Thus $a(x)$ is contained in no divisorial prime ideal. Hence $a(x)$ contains a regular A-sequence f, g. (Suppose $f \in a(x)$, $f \neq 0$. Let $p_1, \ldots, p_r$ be the divisorial prime ideals which contain f. Since $a(x) \not\subseteq p_1 \cup \cdots \cup p_r$, there is a g in $a(x)$ not in any of the $p_1, \ldots, p_r$. Then f, g is a regular A-sequence.) But then $g(fx) = f(gx) = 0$ with fx and gx in M. Hence $fx \in fM$, so $x \in M$. □

Corollary 5.6. *Let A be a Krull domain, M an A-lattice in the vector space V. Let N be an A-submodule of V containing M and set $Q = N/M$.*

(a) If M is divisorial, then $\mathrm{Ass}\, Q \subseteq Z$.

(b) If N is a divisorial A-lattice and $\mathrm{Ann}\, x$ is divisorial for each x in Q, then M is divisorial.

(c) If N is an A-lattice, then $\mathrm{Ass}\, Q \cap Z$ is finite.

(d) *Suppose N and M are divisorial. Then for each $\mathfrak{p}$ in $\mathrm{Ass}\,Q$, there is a divisorial A-lattice $M(\mathfrak{p})$ such that*

   (i) $M \subseteq M(\mathfrak{p}) \subseteq N$

with (ii) $\mathrm{Ass}(N/M(\mathfrak{p})) = \{\mathfrak{p}\}$

and (iii) $M = \bigcap M(\mathfrak{p})$.

   (iv) *The $M(\mathfrak{p})$ are maximal with respect to the properties* (i), (ii) *and* (iii).

(e) *If N and M are divisorial, and $\mathrm{Ass}\,Q = \{\mathfrak{p}\}$, then there is a finite chain of divisorial A-lattices $M = M_r \subsetneqq \cdots \subsetneqq M_0 = N$ such that $\mathfrak{p}M_i \subseteq M_{i+1}$ for $i = 0, \ldots, r-1$ and $\mathrm{Ass}\,M_i/M_{i+1} = \{\mathfrak{p}\}$.*

Proof. (a) and (b) follow from Corollary 5.5. The rest of the proof is straightforward and we leave it for the interested reader. $\square$

These results show that the theory of divisorial lattices is sufficiently rich, in that a theory of associated prime ideals, primary modules and primary decomposition can be developed. Further ramifications of this theory are not needed here. There are many interesting results which can be obtained. Some of them appear in [Bass, 1968], [Beck, 1971], and [Claborn and Fossum, 1968].

We now go on to study divisorial ideals. Let $\mathfrak{a}$ be a nonzero fractionary ideal in K. Let $v_{\mathfrak{p}}$ be the valuation associated to the principal valuation ring $A_{\mathfrak{p}}$. Define $v_{\mathfrak{p}}(\mathfrak{a}) = \inf\{v_{\mathfrak{p}}(a) : a \in \mathfrak{a}\}$. Since $\mathfrak{a}_{\mathfrak{p}} = (\mathfrak{p}A_{\mathfrak{p}})^r$ for some r in $\mathbb{Z}$, it follows that $v_{\mathfrak{p}}(\mathfrak{a}) > -\infty$. In fact $v_{\mathfrak{p}}(\mathfrak{a}) = r$. Also, $v_{\mathfrak{p}}(\mathfrak{a}) = 0$ for almost all $\mathfrak{p}$ in Z.

For $\mathfrak{p}$ in Z, define the *symbolic nth power* of $\mathfrak{p}$ to be the divisorial ideal $\{a \in K : v_{\mathfrak{p}}(a) \geq n\}$ and denote it by $\mathfrak{p}^{(n)}$.

Corollary 5.7. *Let $\mathfrak{a}$ be a fractionary ideal in the quotient field of A. Let $n_{\mathfrak{p}} = v_{\mathfrak{p}}(\mathfrak{a})$. Then $A:(A:\mathfrak{a}) = \bigcap_{\mathfrak{p}} \mathfrak{p}^{(n_{\mathfrak{p}})}$.*

Proof. We get the following sequence of equalities: $A:(A:\mathfrak{a}) = \bigcap \mathfrak{a}_{\mathfrak{p}} = \bigcap (\mathfrak{p}A_{\mathfrak{p}})^{n_{\mathfrak{p}}} = \bigcap \mathfrak{p}^{(n_{\mathfrak{p}})}$. $\square$

Theorem 5.8 (The Approximation Theorem for Krull Domains). *Let A be a Krull domain. Let $n(\mathfrak{p})$ be a given integer for each $\mathfrak{p}$ in Z such that $n(\mathfrak{p}) = 0$ for almost all $\mathfrak{p}$. For any preassigned set $\mathfrak{p}_1, \ldots, \mathfrak{p}_r$ there is an x in K^* such that $v_{\mathfrak{p}_i}(x) = n(\mathfrak{p}_i)$ with $v_{\mathfrak{p}}(x) \geq 0$ otherwise.*

Proof. Let $\mathfrak{p}, \mathfrak{p}_1, \ldots, \mathfrak{p}_r$ be a given finite set of distinct prime ideals in Z. Pick t in $\mathfrak{p}$ such that $tA_{\mathfrak{p}} = \mathfrak{p}A_{\mathfrak{p}}$. Suppose t is in the prime ideals $\mathfrak{p}_1, \ldots, \mathfrak{p}_s$ but is not in $\mathfrak{p}_{s+1}, \ldots, \mathfrak{p}_r$. If $s = 0$, we are finished with our beginning step. Otherwise there are an f_0 in $\mathfrak{p}^2 - \left(\bigcup_1^s \mathfrak{p}_i\right)$ and f_j in $\mathfrak{p}_{s+j} - \left(\bigcup_1^s \mathfrak{p}_i\right)$ for $j > 0$. Then $f = f_0 f_1 \ldots f_{r-s}$ is in $\mathfrak{p}\mathfrak{p}_{s+1} \ldots \mathfrak{p}_r - \left(\bigcup_1^s \mathfrak{p}_i\right)$.

It follows that $t-f$ is in $\mathfrak{p}-(\mathfrak{p}_1\cup\cdots\cup\mathfrak{p}_r)$. As $t\neq f$, and as $f\in\mathfrak{p}^2 A_\mathfrak{p}$ $=t^2 A_\mathfrak{p}$, we get $(t-f)A_\mathfrak{p}=\mathfrak{p}A_\mathfrak{p}$. Hence we have found an element t in $\mathfrak{p}$, but not in $\mathfrak{p}_1\cup\cdots\cup\mathfrak{p}_r$. such that $v_\mathfrak{p}(t)=1$.

Now let $\mathfrak{p}_1,\ldots,\mathfrak{p}_r$ be the preassigned set of distinct prime ideals in Z. As in the paragraph above, pick t_i in A such that $v_{\mathfrak{p}_i}(t_i)=\delta_{ij}$. Let $y=t_1^{n(\mathfrak{p}_1)}\ldots t_r^{n(\mathfrak{p}_r)}$. Then $v_{\mathfrak{p}_j}(y)=n(\mathfrak{p}_j)$ for each j. Let $\mathfrak{q}_1,\ldots,\mathfrak{q}_q$ be those elements of Z distinct from the $\mathfrak{p}_1,\ldots,\mathfrak{p}_r$ such that $v_{\mathfrak{q}_j}(y)<0$. Let m_j be the positive integer such that $m_j=-v_{\mathfrak{q}_j}(y)$. Let s_j, $j=1,\ldots,q$, be elements of A such that $v_{\mathfrak{q}_j}(s_j)=1$ and $v_\mathfrak{p}(s_j)=0$ if $\mathfrak{p}\neq\mathfrak{q}_j$ where $\mathfrak{p}\in\{\mathfrak{p}_1,\ldots,\mathfrak{p}_r,\mathfrak{q}_1,\ldots,\mathfrak{q}_q\}$. Let $x=y s_1^{m_1}\ldots s_q^{m_q}$. Then x satisfies the requirements of the theorem. $\square$

The isomorphism between $D(A)$ and $\mathbb{Z}^{(Z)}$ can now be defined more precisely. Let $\mathfrak{a}$ be a fractionary ideal. Define the divisor of $\mathfrak{a}$, $\operatorname{div}\mathfrak{a}:Z\to\mathbb{Z}$, by $\operatorname{div}(\mathfrak{a})(\mathfrak{p})=v_\mathfrak{p}(\mathfrak{a})$ for each $\mathfrak{p}$ in Z.

Proposition 5.9. *The homomorphism* div *satisfies the properties:*
(a) $\operatorname{div}(A:\mathfrak{a})=-\operatorname{div}\mathfrak{a}$.
(b) $\operatorname{div} A:(A:\mathfrak{a})=\operatorname{div}\mathfrak{a}$.
(c) *If* $\mathfrak{a}\geqslant\mathfrak{b}$, *then* $\operatorname{div}\mathfrak{a}\leqslant\operatorname{div}\mathfrak{b}$.
(d) $\operatorname{div}\mathfrak{a}\mathfrak{b}=\operatorname{div}\mathfrak{a}+\operatorname{div}\mathfrak{b}$.
(e) $\operatorname{div}(\mathfrak{a}+\mathfrak{b})=\inf(\operatorname{div}\mathfrak{a},\operatorname{div}\mathfrak{b})$.
(f) $\operatorname{div}:D(A)\to\mathbb{Z}^{(Z)}$ *is an order preserving isomorphism.* $\square$

Thus any function $n:Z\to\mathbb{Z}$ with finite support is a divisor of some (unique divisorial) fractionary ideal. We call n a *divisor* of A.

Any element x in K^* gives rise to the principal fractionary ideal xA. We obtain a homomorphism $K^* \xrightarrow{\operatorname{div}} \mathbb{Z}^{(Z)}$ defined by $\operatorname{div}(x)=\operatorname{div}(xA)$. These elements are called *principal divisors*.

We denote by $\operatorname{Prin} A$ the subgroup of principal divisors (isomorphic to $P(A)$, the subgroup of principal fractionary ideals) and by $\operatorname{Div} A$ the group of divisors. (This notation is somewhat different from that in [EGA, IV, § 21]. We will see later that the group of invertible ideals is a subgroup of $\operatorname{Div} A$. This subgroup will be denoted by $\operatorname{Cart} A$ for *Cartier divisors*.)

We have the exact sequence

$$1 \longrightarrow A^* \longrightarrow K^* \xrightarrow{\operatorname{div}} \operatorname{Prin} A \longrightarrow 0.$$

Then a restatement of the approximation theorem has the following form.

Corollary 5.10. *Let d be a divisor and Y a finite subset of Z. There is a principal divisor $\operatorname{div}(x)$ such that $\operatorname{div}(x)(\mathfrak{p})=d(\mathfrak{p})$ for all $\mathfrak{p}$ in Y and $\operatorname{div}(x)(\mathfrak{p})\geqslant 0$ otherwise.* $\square$

Let d be a divisor. Let $Y' = \mathrm{Supp}(d)$. Then $Y' = \{\mathfrak{p} \in Z : d(\mathfrak{p}) \neq 0\}$. Let x in K^* be an element such that $\mathrm{div}(x) = d$ on Y'. Set $Y'' = \mathrm{Supp}(\mathrm{div}\, x) - Y'$. The set Y'' is finite. Find y in K^* such that $\mathrm{div}(y)(\mathfrak{p}) = d(\mathfrak{p})$ for all $\mathfrak{p}$ in $\mathrm{Supp}(d)$, $\mathrm{div}(y)(\mathfrak{p}) = 0$ for all $\mathfrak{p}$ in Y'' and $\mathrm{div}(y)(\mathfrak{p}) \geqslant 0$ otherwise. It is then easy to see that

$$\inf(\mathrm{div}(x), \mathrm{div}(y)) = d.$$

However $\inf(\mathrm{div}(x), \mathrm{div}(y)) = \mathrm{div}(A\,x + A\,y)$ according to Proposition 5.9e. We restate this result in terms of ideals.

Proposition 5.11. *Let $\mathfrak{a}$ be a fractionary ideal. There are elements x, y in $\mathfrak{a}$ such that $A : (A : \mathfrak{a}) = A : (A : (Ax + Ay))$.* $\square$

For example, when every ideal is divisorial, a situation which attains if and only if A is a Dedekind domain, then every ideal is generated by two elements.

Chapter II. The Divisor Class Group and Factorial Rings

In this chapter the divisor class group of a Krull domain is introduced and its functorial properties are investigated. Two applications of the functorial properties are made to polynomial extensions and subintersections (cf. Proposition 1.2d). Power series extensions are considered in a later section.

All rings in this chapter are Krull domains (unless mentioned otherwise). Once again let $Z(A) = X^{(1)}(A)$. When no confusion can arise, this is denoted by Z.

§ 6. The Divisor Class Group and its Functorial Properties

Let A be a Krull domain. Recall the isomorphism $\mathrm{div}: D(A) \to \mathrm{Div}(A)$ from the group of fractionary divisorial ideals to the free group on the set of minimal nonzero prime ideals. For $0 \neq x$ in the field of quotients of A, let $\mathrm{div}(x)$ denote $\mathrm{div}(Ax)$. The set of principal divisorial fractionary ideals forms a subgroup $\mathrm{Prin}(A)$ of $D(A)$. We denote also by $\mathrm{Prin}(A)$ the image in $\mathrm{Div}(A)$ of $\mathrm{Prin}(A)$.

The *divisor class group* of the Krull domain A is the group $\mathrm{Div}(A)/\mathrm{Prin}(A)$. It is denoted by $\mathrm{Cl}(A)$. For the fractionary ideal $\mathfrak{a}$ we denote by $\mathrm{cl}(\mathfrak{a})$ the image in $\mathrm{Cl}(A)$ of $\mathrm{div}(\mathfrak{a})$.

Recall that an integral domain A is a *factorial* ring (or a *unique factorization domain* or a *UFD*) provided every element in A is uniquely (up to multiplication by a unit) a finite product of irreducible (or prime) elements. A factorial ring is a Krull domain.

The next proposition shows that $\mathrm{Cl}(A)$ measures, to some extent, the lack of unique factorization in A.

Proposition 6.1. *Let A be an integral domain. The following are equivalent statements.*

(i) *A is a factorial ring.*

(ii) *Every nonzero prime ideal of A contains a nonzero principal prime ideal.*

(iii) *A is a Krull domain with* $Cl(A) = (0)$.

(iv) *A is a Krull domain in which each* $\mathfrak{p}$ *in* $X^{(1)}(A)$ *is principal.*

(v) *A is an integral domain which satisfies the ascending chain condition for principal ideals and the intersection of any two principal ideals is principal* (and generated by the least common multiple).

(vi) *A has the ascending chain condition for principal ideals and a maximal principal ideal is a prime ideal.*

(vii) *There is a function* $s: A^* \to \mathbb{N}$ *such that:*

(a) $s(xy) = s(x) + s(y)$.

(b) $s(x) = 0$ *if and only if x is a unit in A.*

(c) *If* $x, y \in A^*$ *with* $xA \nsubseteq yA$, $yA \nsubseteq xA$, *then there are* a,b,c,d *in* A^* *such that* $ax + by = cd$ *with* $s(c) < s(x)$ *and with* d,x *and* d,y *both regular A-sequences* (i.e., $Ad \cap Ax = Adx$ *and* $Ad \cap Ay = Ady$).

Proof. That (ii) is equivalent to (i) is due to Kaplansky [1970]. The equivalence of (vii) to (i) is found in Bourbaki [1965, 3, Exercise 1]. The other equivalences are also found in Bourbaki.   □

Let A and B be Krull domains with A contained in B. Suppose $\mathfrak{P} \in X^{(1)}(B)$ and $\mathfrak{P} \cap A \in X^{(1)}(A)$. Denote by $\mathfrak{p}$ the ideal $\mathfrak{P} \cap A$. The ramification index of $\mathfrak{p}$ in $\mathfrak{P}$ is the unique integer e such that $\mathfrak{p} B_{\mathfrak{P}} = \mathfrak{P}^e B_{\mathfrak{P}}$. Denote this integer by $e(\mathfrak{P}/\mathfrak{p})$. The assignments

$$\mathfrak{p} \mapsto \sum e(\mathfrak{P}/\mathfrak{p}) \operatorname{div} \mathfrak{P}, \qquad \mathfrak{P} \cap A = \mathfrak{p}, \mathfrak{P} \in X^{(1)}(B)$$

define a homomorphism $\operatorname{Div} i: \operatorname{Div}(A) \to \operatorname{Div}(B)$. As is seen in [Samuel, 1964, Theorem 6.1], this induces a homomorphism $Cl(A) \to Cl(B)$ if and only if the condition

(PDE) "*If* $\mathfrak{P} \in X^{(1)}(B)$, *then* $\operatorname{ht}(\mathfrak{P} \cap A) \leqslant 1$."

is satisfied.[1]

Since $B_{\mathfrak{P}}$ is an $A_{\mathfrak{p}}$-algebra and $A_{\mathfrak{p}}$ is a principal valuation ring provided condition (PDE) is satisfied, we see that $B_{\mathfrak{P}}$ is a faithfully flat $A_{\mathfrak{p}}$-module. The converse holds as well. That is, if $B_{\mathfrak{P}}$ is a (faithfully) flat $A_{\mathfrak{p}}$-module, then $\dim A_{\mathfrak{p}} \leqslant \dim B_{\mathfrak{P}}$ and so $\operatorname{ht} \mathfrak{p} \leqslant 1$.

Theorem 6.2. *Suppose A and B are Krull domains and* $A \hookrightarrow B$. *Let* $\operatorname{Div} i: \operatorname{Div} A \to \operatorname{Div} B$ *be the induced homomorphism. The following statements are equivalent:*

---

[1] PDE is the abbreviation for *pas d'éclatement* or *no blowing up* (NBU) as in [Samuel, loc. cit.].

(i) *The image* $\mathrm{Div}(i)(\mathrm{Prin}(A)) \subseteq \mathrm{Prin}(B)$.

(ii) *Condition* (PDE) *is satisfied.*

(iii) *If* $\mathfrak{P} \in X^{(1)}(B)$, *then* $B_{\mathfrak{P}}$ *is a flat* A-*module.*

(iv) *The restriction of* $^a i : \mathrm{Spec}\, B \to \mathrm{Spec}\, A$ *to the subset* $X^{(1)}(B)$ *has its image in* $X^{(1)}(A) \cup \{(0)\}$.

(v) *For all factionary ideals* $\mathfrak{a}, \mathfrak{b}$ *of* A, *the equality* $A : \mathfrak{a} = A : \mathfrak{b}$ *implies* $B : \mathfrak{a} B = B : \mathfrak{b} B$.

Proof. That (i) and (ii) are equivalent is found in [Samuel, 1964d, Theorem 6.1]. Statement (iv) is equivalent to (ii). It was mentioned above that (iii) is equivalent to (ii). That (v) is equivalent to (ii) is left to the interested reader. (See also [EGA, IV, §21].)

It is the functorial properties of the divisor class group which will be studied in this section. Later, in Section 18, we will see that a map on the divisor class groups is sometimes, but rarely, induced by a map from a ring onto another ring. But this is not induced by the restriction $^a i : X^{(1)}(B) \to X^{(1)}(A)$.

However, in case condition (PDE) is satisfied, the associated homomorphisms have very nice properties.

Proposition 6.3. *Suppose the* A-*algebra* B *satisfies condition* (PDE). *Then* $\mathrm{Div}(i)$ *has the following properties:*

(a) *For all* x *in* K^*, $\mathrm{Div}(i)(\mathrm{div}_A(x)) = \mathrm{div}_B(x)$.

(b) *The homomorphism* $\mathrm{Div}(i)$ *is order preserving.*

(c) *The homomorphism* $\mathrm{Div}(i)$ *preserves least common multiples. That is,* $\mathrm{Div}(i)(\mathrm{l.c.m.}(\mathrm{div}(\mathfrak{a}), \mathrm{div}(\mathfrak{b}))) = \mathrm{l.c.m.}(\mathrm{Div}(i)(\mathrm{div}(\mathfrak{a})), \mathrm{Div}(i)(\mathrm{div}(\mathfrak{b})))$.

(d) *In terms of ideals in* $\mathrm{D}(A)$, *the homomorphism is defined by* $\mathrm{Div}(i)(\mathfrak{a}) = B : (B : \mathfrak{a} B)$. $\quad\square$

The least common multiple of two divisors is defined to be the function whose values are given by $\mathrm{l.c.m.}(\mathrm{div}(\mathfrak{a}), \mathrm{div}(\mathfrak{b}))(\mathfrak{p}) = \sup\{\mathrm{div}(\mathfrak{a})(\mathfrak{p}), \mathrm{div}(\mathfrak{b})(\mathfrak{p})\}$. The greatest common divisor is defined using inf.

There are several general cases in which there is no blowing up for divisorial ideals.

Proposition 6.4. *In any of the cases below, the* A-*algebra* B *satisfies condition* (PDE).

(a) *The* A-*algebra* B *is flat as an* A-*module.*

(b) *The* A-*algebra* B *is integral over* A.

(c) *The* A-*algebra* B *is a subintersection.* $\quad\square$

In case B is a flat A-module, the extension $\mathfrak{a} B$ of a divisorial ideal $\mathfrak{a}$ of A is again divisorial.

We note that none of these conditions overlap. Certainly not every flat extension is an integral extension, nor is it a subintersection. Integral extensions need not be flat. A subintersection is integral if and only if it is not proper. We need to see that not every subintersection is flat. We now show that a flat extension of a Krull domain inside its field of quotients is a subintersection but not conversely. We use the following useful lemma.

Lemma 6.5. *Suppose A is an integral domain whose field of quotients is K. Let B be a ring between A and K. The following statements are equivalent:*

(i) *The ring B is flat as an A-module.*

(ii) *For all x in K^*, the ideal $B:(B+Bx) = B(A:A+Ax)$.*

(iii) *For all maximal ideals $\mathfrak{M}$ in B, the localization $B_{\mathfrak{M}} = A_{\mathfrak{M} \cap A}$.*

Proof. Use *la formule des transporteurs* of [Bourbaki, 1961, page 41] to see that (i) implies (ii).

Suppose (ii). Let x be an element in $B_{\mathfrak{M}}$. We seek an element a in $A - (\mathfrak{M} \cap A)$ such that $ax \in A$. Write $x = b/s$ with b in B and s in $B - \mathfrak{M}$. Then $s \in B:(B+Bx)$. Hence there are elements b_i in B and w_i in $A:A+Ax$ such that $s = \sum_{i=1}^{n} b_i w_i$. Since $s \notin \mathfrak{M}$, at least one $w_i \notin \mathfrak{M} \cap A$. But $w_i x \in A$. Hence (ii) implies (iii).

The implication (iii) implies (i) is the local characterization of flatness.   ∎

The only novelty of this lemma is condition (ii) which was noticed by Istvan Beck.

Corollary 6.6. *A flat extension of a Krull domain within its field of quotients is a subintersection.*

Proof. Let B be a flat extension of the Krull domain A. Then
$$B = \bigcap_{\mathfrak{M}} B_{\mathfrak{M}}.$$
By the lemma, the intersection $\bigcap_{\mathfrak{M}} B_{\mathfrak{M}} = \bigcap_{\mathfrak{M}} A_{\mathfrak{M} \cap A}$. But each $A_{\mathfrak{M} \cap A}$ is a subintersection, so therefore B is a subintersection.   ∎

Consequently we can give an example of a non-flat subintersection of a noetherian Krull domain, thereby answering a question first posed by Beck.

Consider the Krull domain $k[x,y,u,v]$ with $xv = yu$ defined over the field k, which is discussed later in § 14. The local ring $k[x,y,u,v]_{(x,y,u,v)}$ is not regular. The ideal $\mathfrak{M}$ generated by x, y and u/x in the extension $k[x,y,u/x]$ of $k[x,y,u,v]$ is maximal. It meets $k[x,y,u,v]$ in the maximal ideal (x,y,u,v). But $k[x,y,u/x]$ is a pure transcendental extension of k. By the Hilbert syzygy theorem, the localization $k[x,y,u/x]_{\mathfrak{M}}$ is a regular local ring. Hence $k[x,y,u/x]_{\mathfrak{M}} \neq k[x,y,u,v]_{(x,y,u,v)}$. So $k[x,y,u/x]$ is not a flat $k[x,y,u,v]$-module. But $k[x,y,u,v] = k[x,y,u/x] \cap k[x,y,u,v]_{(x,y)}$. So $k[x,y,u/x]$ is a subintersection.

Special cases in which subintersections are flat do arise.

Proposition 6.7. *Every subintersection of the Krull domain A is a ring of quotients* (and so flat) *if and only if $\mathrm{Cl}(A)$ is a torsion group.*

Proof. Suppose every subintersection is a ring of quotients. To show that $\mathrm{Cl}(A)$ is torsion, it is sufficient to show that $\mathrm{cl}(\mathfrak{p})$ is a torsion element for every $\mathfrak{p}$ in $X^{(1)}(A)$. Let $\mathfrak{p}$ be such an element. Set $B = \bigcap_{\mathfrak{q} \neq \mathfrak{p}} A_{\mathfrak{q}}$. Then B is a subintersection so there is a multiplicatively closed subset S of A such that $B = S^{-1}A$. Now $B = B:(B:\mathfrak{p}B)$. But since B is flat, the ideal $\mathfrak{p}B$ is divisorial. Hence $\mathfrak{p}B = B$. Then $S \cap \mathfrak{p} \neq \emptyset$. For those divisorial prime ideals $\mathfrak{q}$ different from $\mathfrak{p}$ we have $\mathfrak{q}B \neq B$. Hence $\mathfrak{q} \cap S = \emptyset$. Suppose $s \in S \cap \mathfrak{p}$. Then the principal ideal sA is contained in only $\mathfrak{p}$ and is therefore a symbolic power of $\mathfrak{p}$, say $sA = \mathfrak{p}^{(n)}$. Hence $n \cdot \mathrm{cl}(\mathfrak{p}) = 0$. So $\mathrm{cl}(\mathfrak{p})$ is torsion.

Suppose now that $\mathrm{Cl}(A)$ is a torsion group. Let B be a subintersection, say $B = \bigcap_{\mathfrak{q} \in T} A_{\mathfrak{q}}$ where $T \subseteq X^{(1)}(A)$. For each $\mathfrak{p}$ not in T, let $s_{\mathfrak{p}}$ be an element in $\mathfrak{p}$ such that $\mathfrak{p}^{(n)} = s_{\mathfrak{p}}A$ for a suitable $n > 0$. Let S be the multiplicative subset generated by these elements $s_{\mathfrak{p}}$. Then $B = S^{-1}A$, so each subintersection is a ring of quotients. □

Storch [1967] calls a Krull domain with a torsion divisor class group an *almost factorial ring* (*fast faktoriell*). Storch has the next result.

Proposition 6.8. *The following statements are equivalent:*

(i) *The divisor class group of A is a torsion group.*

(ii) *Every subintersection of A is a ring of quotients.*

(iii) *Some power of each nonzero nonunit element in A is a product of primary elements.*

(iv) *If f and g are elements in A, then there is an integer $n > 0$ (depending on f and g) such that*

$$Af^n \cap Ag^n$$

is a principal ideal.

(v) *If $\mathfrak{p} \in X^{(1)}(A)$, then $\mathfrak{p}^{(n)}$ is principal for some $n > 0$.*

Proof. Assuming the equivalence of (i) and (v) we have demonstrated the equivalence of (i) and (ii) in the proof of Proposition 6.7.

Suppose (v). Let f be an element of A. Then the ideal fA has the primary decomposition $\mathfrak{p}_1^{(e_1)} \cap \cdots \cap \mathfrak{p}_r^{(e_r)} = fA$. Then $f^n A = \mathfrak{p}_1^{(e_1 n)} \cap \cdots \cap \mathfrak{p}_r^{(e_r n)}$. But $\mathfrak{p}_j^{(e_j n)}$ is principal for each j provided n is chosen suitably. So $f^n A = g_1 \ldots g_r A$ for primary elements $g_1, \ldots, g_r$. Hence (v) implies (iii).

A straightforward calculation shows that (iii) implies (iv). So there remains the verification that (iv) implies (v).

Suppose (iv) and let $\mathfrak{p}$ be a prime divisorial ideal. By the approximation theorem Proposition 5.11, there are elements f and g in $\mathfrak{p}$ such that $v_{\mathfrak{p}}(f)=1$ and $v_{\mathfrak{p}}(g)=1$ while $v_{\mathfrak{q}}(f)v_{\mathfrak{q}}(g)=0$ for all $\mathfrak{q}\neq\mathfrak{p}$. By (iv), there is an $n>0$ such that $Af^n\cap Ag^n$ is principal, say $Af^n\cap Ag^n=Ah$. Let e denote $(fg)^n/h$. Then $e\in A$ and $v_{\mathfrak{p}}(e)+v_{\mathfrak{p}}(h)=n(v_{\mathfrak{p}}(f)+v_{\mathfrak{p}}(g))$. But $v_{\mathfrak{p}}(h)=n$. Hence $v_{\mathfrak{p}}(e)=n$. If $\mathfrak{q}\neq\mathfrak{p}$, then $v_{\mathfrak{q}}(e)+v_{\mathfrak{q}}(h)=n(v_{\mathfrak{q}}(f)+v_{\mathfrak{q}}(g))$, whereas $v_{\mathfrak{q}}(h)=n(v_{\mathfrak{q}}(f)+v_{\mathfrak{q}}(g))$. Hence $v_{\mathfrak{q}}(e)=0$. Thus $eA=\mathfrak{p}^{(n)}$.   □

Since the class number of a ring of algebraic integers is finite, these rings provide a large class of examples of almost factorial rings.

Storch notes that a noetherian integral domain which satisfies (v) is normal (and hence an almost factorial Krull domain).

A further comment concerning the order of elements in $\mathrm{Cl}(A)$ when A is almost factorial is needed. It follows directly from the proof.

Proposition 6.9. *Let A be an almost factorial integral domain. If the order of $\mathrm{Cl}(A)$ divides n then $f^nA\cap g^nA$ is principal for all f,g in A, and conversely.*   □

We have seen that a flat extension of Krull domains $A\to B$ induces a group homomorphism $\mathrm{Cl}(A)\to\mathrm{Cl}(B)$. Under certain nicer conditions, as is seen in the next section, this is an epimorphism. There is a general situation when this can be seen to be a monomorphism.

Recall that the extension $A\to B$ is faithfully flat if it is flat and if, for any maximal ideal $\mathfrak{m}$ of A, $\mathfrak{m}B\neq B$.

Proposition 6.10. *Let $A\to B$ be a faithfully flat extension.*

(a) If B is a Krull domain then A is a Krull domain.

(b) If $\mathfrak{a}$ is an ideal in A such that $\mathfrak{a}B$ is principal, then $\mathfrak{a}$ is invertible (i.e., $(A:\mathfrak{a})\mathfrak{a}=A$).

In particular, if invertible ideals in A are principal or if the extended ideal $\mathfrak{a}B$ can be generated by an element in $\mathfrak{a}$, then $\mathrm{Cl}(A)\to\mathrm{Cl}(B)$ is a monomorphism.

Proof. (a) Let K denote the field of quotients of A. Suppose $a/b\in K\cap B$. Then $\{f\in A:fa\in bA\}B=B$. Since $A\to B$ is faithfully flat, the two ideals $\{f\in A:fa\in bA\}$ and $A\cap\{f\in A:fa\in bA\}B$ are equal. Hence $a/b\in A$. So A is a Krull domain by Proposition 1.2.

(b) Since faithfully flat ring homomorphisms reflect finite generation and projectivity, the assumption that $\mathfrak{a}B$ is principal implies that $\mathfrak{a}$ is finitely generated and projective. Hence $\mathfrak{a}$ is invertible.

The last statement then follows.   □

In Section 18 we will meet the Picard group of invertible ideals. The Picard group $\mathrm{Pic}\,A$ is the subgroup of $\mathrm{Cl}(A)$ of classes of invertible ideals. Proposition 6.10(b) can be stated that $\mathrm{Ker}(\mathrm{Cl}(A) \to \mathrm{Cl}(B))$ $= \mathrm{Ker}(\mathrm{Pic}(A) \to \mathrm{Pic}(B))$. Thus in cases that $\mathrm{Pic}(A) \to \mathrm{Pic}(B)$ is a monomorphism, for example when B is a local ring, then $\mathrm{Cl}(A) \to \mathrm{Cl}(B)$ is also a monomorphism.

Corollary 6.11. *If B is a local (noetherian) Krull domain and if $A \to B$ is a faithfully flat extension, then $\mathrm{Cl}(A) \to \mathrm{Cl}(B)$ is a monomorphism.*

Proof. A is then local, and so invertible ideals are principal. □

Corollary 6.12 (Mori's Theorem). *Let $A, \mathfrak{m}$ be a Zariski pair (i.e., $\mathfrak{m}$ is contained in the Jacobson radical of A). If the $\mathfrak{m}$-adic completion $\hat{A}$ is normal, then A is normal and*

$$\mathrm{Cl}(A) \to \mathrm{Cl}(\hat{A})$$

is a monomorphism. In particular, if $\hat{A}$ is factorial (almost factorial) then A is factorial (almost factorial). □

The extension $A \to A[[X]]$ is also faithfully flat when A is noetherian. If $\mathfrak{a}$ is an ideal in A such that $\mathfrak{a}A[[X]]$ is principal, then, again by Nakayama's Lemma, there is an element f in $\mathfrak{a}$ such that $\mathfrak{a}A[[X]]$ $= fA[[X]]$. But then $\mathfrak{a}$ is principal. Hence we can conclude that $\mathrm{Cl}(A) \to \mathrm{Cl}(A[[X]])$ is a monomorphism.

Corollary 6.13. *Let A be a noetherian normal integral domain. Then*

$$\mathrm{Cl}(A) \to \mathrm{Cl}(A[[X]])$$

is a monomorphism. □

We devote most of Chapter V to the study of the cokernel of the monomorphisms $\mathrm{Cl}(A) \to \mathrm{Cl}(\hat{A})$ and $\mathrm{Cl}(A) \to \mathrm{Cl}(A[[X]])$. In Section 17 we see that these are not necessarily isomorphisms.

§ 7. Nagata's Theorem

We wish to study, in this section, the map $\mathrm{Cl}(A) \to \mathrm{Cl}(B)$, where B is a subintersection. Since each ring of quotients is a subintersection, we thus will also get information about the map $\mathrm{Cl}(A) \to \mathrm{Cl}(S^{-1}A)$ where S is a multiplicatively closed subset of A. The first result is called Nagata's theorem, since it is a straightforward generalization of his theorem concerning the homomorphism $\mathrm{Cl}(A) \to \mathrm{Cl}(S^{-1}A)$. One is referred to Nagata [1957] or Samuel [1964c].

Theorem 7.1 (Nagata's Theorem). *Let A be a Krull domain and suppose $B = \bigcap_{\mathfrak{q} \in T} A_{\mathfrak{q}}$, a subintersection for A with $T \subseteq X^{(1)}(A)$. Then*

(i) $\mathrm{Cl}(A) \to \mathrm{Cl}(B)$ *is a surjection.*

(ii) $\mathrm{Ker}(\mathrm{Cl}(A) \to \mathrm{Cl}(B))$ *is generated by the classes of those prime ideals not in* T.

Proof. The homomorphism $\mathrm{Cl}(A) \to \mathrm{Cl}(B)$ is induced from the homomorphism $\mathrm{Div}\,A \to \mathrm{Div}\,B$. But $X^{(1)}(B) \cong T \subseteq X^{(1)}(A)$, the surjection being given by $\mathfrak{P} \mapsto \mathfrak{P} \cap A$ with inverse $\mathfrak{p} \mapsto (\mathfrak{p}B)^{**}$ for $\mathfrak{p}$ in T. Hence $\mathrm{Div}\,A \to \mathrm{Div}\,B$ is a surjection, which implies that $\mathrm{Cl}(A) \to \mathrm{Cl}(B)$ is a surjection. We now show that the image in $\mathrm{Cl}(A)$ of the kernel of $\mathrm{Div}\,A \to \mathrm{Div}\,B$ is exactly the kernel of the homomorphism $\mathrm{Cl}(A) \to \mathrm{Cl}(B)$. To see this, it is sufficient (and necessary) to show that

$$\mathrm{Prin}(A) \to \mathrm{Prin}(B)$$

is a surjection. But $\mathrm{Prin}(B) = \mathrm{im}(K^* \to \mathrm{Div}\,B)$ and $\mathrm{Prin}(A) = \mathrm{im}(K^* \to \mathrm{Div}\,A)$. Furthermore,

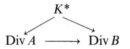

is commutative. Hence $\mathrm{Prin}\,A \to \mathrm{Prin}\,B$ is a surjection. □

Corollary 7.2. *Let S be a multiplicatively closed subset of A. Then*

(i) $\mathrm{Cl}(A) \to \mathrm{Cl}(S^{-1}A)$ *is a surjection.*

(ii) *The kernel is generated by the classes of the prime ideals which meet S.* □

Corollary 7.3. *Let S be a multiplicatively closed subset of A. If S is generated by prime elements, then $\mathrm{Cl}(A) \to \mathrm{Cl}(S^{-1}A)$ is a bijection. In particular, if $S^{-1}A$ is factorial, then A is factorial.* □

These two corollaries will be used extensively in the next few sections, where applications of Nagata's Theorem will allow us to find many factorial rings, and compute the divisor class groups of many others.

§ 8. Polynomial Extensions

In Section 1 we saw that $A[X]$ is a Krull domain if and only if A is a Krull domain. Using Nagata's theorem, we will show that if A is factorial then so is $A[X]$, the well known application of Gauss' Lemma.

Theorem 8.1. *Let A be a Krull domain. Then the induced homomorphism*

$$\mathrm{Cl}(A) \to \mathrm{Cl}(A[X])$$

is a bijection.

Proof. Denote by S the nonzero elements in A. Then $S^{-1}A[X]$ $= K[X]$. But $K[X]$ is a principal ideal ring and is therefore factorial. By Nagata's result, the group $\mathrm{Cl}(A[X])$ is generated by the prime ideals which meet S. But if $\mathfrak{P} \in X^{(1)}(A[X])$ and $\mathfrak{P} \cap A \neq 0$, then $\mathfrak{P} = (\mathfrak{P} \cap A)A[X]$. Hence $\mathrm{Cl}(A) \to \mathrm{Cl}(A[X])$ is a surjection.

Suppose that M is an A-module. Then the natural homomorphism $M \to A \otimes_{A[X]} A[X] \otimes_A M$ is a bijection. Thus if $\mathfrak{a}$ is an ideal in A such that $\mathfrak{a}A[X]$ is principal, then $\mathfrak{a}$ is principal. Therefore $\mathrm{Cl}(A) \to \mathrm{Cl}(A[X])$ is an injection. $\square$

Corollary 8.2. *The Krull domain A is factorial if and only if $A[X]$ is factorial.* $\square$

Suppose X is a set of indeterminants. For an element f in $K[X]$, define the content ideal $\mathfrak{C}(f)$ of f to be the fractionary ideal in K generated by the coefficients of f. If A is factorial, then $\mathrm{div}\,\mathfrak{C}(f)$ is a principal divisor generated by the greatest common divisor of the coefficients of f. Call this element $c(f)$.

The lemma of Dedekind-Mertens [Gilmer, 1967] gives a relation between the content ideals $\mathfrak{C}(f)$, $\mathfrak{C}(g)$ and $\mathfrak{C}(fg)$.

Lemma 8.3. *Let f and g be nonzero elements in $K[X]$. Then there is an integer $k > 0$ such that $\mathfrak{C}(f)^{k+1}\mathfrak{C}(g) = \mathfrak{C}(f)^k\mathfrak{C}(fg)$.* $\square$

Corollary 8.4. *Suppose A is a Krull domain. If f and g are nonzero elements in $K[X]$, then*

$$\mathrm{div}\,(\mathfrak{C}(f)) + \mathrm{div}\,(\mathfrak{C}(g)) = \mathrm{div}\,(\mathfrak{C}(fg)).   \quad \square$$

Corollary 8.5 (Gauss' Lemma). *Let A be a factorial ring. Then*

$$c(f)c(g) = c(fg).   \quad \square$$

By induction on the number of variables it follows from Corollary 8.2 that $\mathrm{Cl}(A) \to \mathrm{Cl}(A[X_1, \ldots, X_n])$ is a bijection. In fact this is the case for any set of indeterminates. One can prove this by observing that a polynomial involves only a finite number of the indeterminates. However, there is a more general result, which is interesting in itself, and which gives the desired conclusion. It is found in [Gilmer and Heinzer, 1968].

Proposition 8.6. *Let $\{A_\alpha\}_{\alpha \in I}$ be a filtered family of Krull domains such that for $\alpha < \beta$, the extension $A_\alpha \to A_\beta$ satisfies condition (PDE) and $(\mathfrak{p}A_\beta)^{**}$ is a divisorial prime ideal for each $\mathfrak{p}$ in $X^{(1)}(A_\alpha)$. Then $\bigcap A_\alpha$ and $\bigcup_{\alpha \in I} A_\alpha$ are Krull domains. Furthermore, $X^{(1)}\left(\bigcup_{\alpha \in I} A_\alpha\right) = \bigcup_{\alpha \in I} X^{(1)}(A_\alpha)$ and there is induced a bijection*

$$\varinjlim_\alpha \mathrm{Cl}(A_\alpha) \longrightarrow \mathrm{Cl}\left(\bigcup_\alpha A_\alpha\right).$$

Proof. The verification that $\bigcap A_\alpha$ is a Krull domain is left to the interested reader. Let $B = \bigcup A_\alpha$.

Let K_α denote the field of quotients of A_α for each α in I. Then the field of quotients of $\bigcup_\alpha A_\alpha$ is $\bigcup_{\alpha \in I} K_\alpha$, which is denoted by K.

For a fixed α in I, let $I_\alpha = \{\beta \in I : \beta \geq \alpha\}$. Let B_α denote the ring $\bigcup_{\beta \in I_\alpha} A_\beta$. Then $B = B_\alpha$ for each α. Suppose $\mathfrak{p} \in X^{(1)}(A_\alpha)$. For each $\beta \geq \alpha$, there is a unique prime ideal $\mathfrak{P}_\beta$ in $X^{(1)}(A_\beta)$ such that $\mathfrak{P}_\beta \cap A_\alpha = \mathfrak{p}$. Then $(A_\beta)_{\mathfrak{P}_\beta}$ contains $(A_\alpha)_\mathfrak{p}$. Let $V_\mathfrak{p}$ be the union $\bigcup_{\beta \geq \alpha} (A_\beta)_{\mathfrak{P}_\beta}$. This is a principal valuation ring in K. Since it is the union of valuation rings, it is a valuation ring of rank one. Since $\mathfrak{p}(A_\beta)_{\mathfrak{P}_\beta}$ is the maximal ideal of $(A_\beta)_{\mathfrak{P}_\beta}$, it generates the maximal ideal of $V_\mathfrak{p}$. So the valuation ring is principal.

Now $B = \bigcap_{\alpha, \mathfrak{p} \in X^{(1)}(A_\alpha)} V_\mathfrak{p}$. For certainly $B \subseteq V_\mathfrak{p}$ for each such $\mathfrak{p}$. On the other hand, if $x \in \bigcap V_\mathfrak{p}$, then $x \in K_\beta \cap (\bigcap V_\mathfrak{p})$ for some β in I. But $K_\beta \cap (\bigcap_\mathfrak{p} V_\mathfrak{p}) = \bigcap_\mathfrak{p} (K_\beta \cap V_\mathfrak{p})$, and $\bigcap_\mathfrak{p} (K_\beta \cap V_\mathfrak{p}) = A_\beta$. Hence $x \in A_\beta$ for some $\beta \in I$, so $x \in B$.

It is now clear that $\bigcup_\alpha X^{(1)}(A_\alpha) = X^{(1)}(B)$ and that $A_\alpha \to B$ satisfies condition (PDE). Since $\mathrm{Div}(B) = \varinjlim_\alpha \mathrm{Div}(A_\alpha)$, the same holds for the divisor class group. ☐

Suppose X denotes a set of indeterminates over a ring A, say $X = \{X_i\}_{i \in I}$. Let F be a (finite) subset of I. Denote by $A[F]$ the ring of polynomials $A[\{X_i\}_{i \in F}]$. If $F \subseteq F'$, then the natural injection $A[F] \to A[F']$ satisfies condition (PDE). Furthermore if $\mathfrak{p}$ is a prime ideal in $A[F]$, then $\mathfrak{p}A[F']$ is a prime ideal in $A[F']$. Thus the conditions of Proposition 8.6 are fulfilled for the family $\{A[F]\}_{F \subseteq I, F \text{ finite}}$.

Corollary 8.7. *Suppose A is a factorial ring. Then $A[X]$ is factorial.*

Proof. The polynomial ring $A[X] = \bigcup_{F \subseteq I} A[F]$ and the divisor class group $\mathrm{Cl}(A[X]) = \varinjlim_F \mathrm{Cl}(A[F])$. But each $A[F]$ is factorial, so $A[X]$ is factorial. ☐

Proposition 8.8. *Suppose A is a Krull domain. Then $A[X]$ is a Krull domain and the homomorphism $\mathrm{Cl}(A) \to \mathrm{Cl}(A[X])$ is a bijection.*

Proof. From Proposition 8.6, the ring $A[X]$ is a Krull domain and $\mathrm{Cl}(A[X]) = \varinjlim_F \mathrm{Cl}(A[F])$. Since

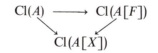

is commutative and since each $\mathrm{Cl}(A) \to \mathrm{Cl}(A[F])$ is a bijection, the homomorphism $\mathrm{Cl}(A) \to \mathrm{Cl}(A[X])$ is a bijection. $\square$

In the polynomial ring $A[X_1, ..., X_r]$ let S be the multiplicatively closed set $\{f : \mathfrak{C}(f) = A\}$. Then $S = \{f \in A[X_1, ..., X_r]$: coefficients of f generate the unit ideal$\}$. It is multiplicatively closed by the Dedekind-Mertens lemma. Denote by $A(X_1, ..., X_r)$ the ring $S^{-1}A[X_1, ..., X_r]$. The extension $A \to A(X_1, ..., X_r)$ is faithfully flat [Nagata, 1962].

Proposition 8.9. *Let A be a local Krull domain. Then*

$$\mathrm{Cl}(A) \to \mathrm{Cl}(A(X_1, ..., X_r))$$

is a bijection.

Proof. The homomorphism $\mathrm{Cl}(A[X_1, ..., X_r]) \to \mathrm{Cl}(A(X_1, ..., X_r))$ is a surjection. Hence the composition $\mathrm{Cl}(A) \to \mathrm{Cl}(A(X_1, ..., X_r))$ is a surjection by Nagata's Theorem 7.1. By Corollary 6.11, the homomorphism is an injection. $\square$

§ 9. Regular Local Rings

In this section, Nagata's theorem is combined with the Auslander-Buchsbaum-Serre homological characterization of regular local rings to show that regular local rings are factorial. This is essentially Kaplansky's proof of the theorem.

Theorem 9.1 (Auslander-Buchsbaum). *If A is a regular local ring, then A is factorial.*

Proof. The proof goes by induction on $\dim A$. If $\dim A = 0$, then A is a field, so it is factorial. If $\dim A = 1$. Then A is a principal valuation ring. Hence A is factorial.

Suppose we have shown that all regular local rings with dimension at most $n-1$, $n \geq 1$, are factorial. Let A be a regular local ring with $\dim A = n$. There is an element f in $\mathfrak{m} - \mathfrak{m}^2$. It follows that f is a prime element (since A/fA is a regular local ring). Hence, by Nagata's Theorem, the homomorphism $\mathrm{Cl}(A) \to \mathrm{Cl}(A_f)$ is a bijection. We now show that A_f is factorial. It is sufficient to show that each divisorial ideal in A_f is principal.

Let $\mathfrak{b}$ be a divisorial ideal in A_f. Then $\mathfrak{b} = \mathfrak{a} A_f$ where $\mathfrak{a}$ is a divisorial ideal in A. Let $\mathfrak{N}$ be a maximal ideal in A_f. Then $(A_f)_{\mathfrak{N}} = A_{\mathfrak{n}}$ where $\mathfrak{n} = A \cap \mathfrak{N}$. Hence $(A_f)_{\mathfrak{N}}$ is a regular local ring with $\dim(A_f)_{\mathfrak{N}} < \dim A$. Thus $\mathfrak{b}(A_f)_{\mathfrak{N}}$, being a divisorial ideal, is principal. Hence $\mathfrak{b}$ is a projective ideal in A_f.

Since A is a regular local ring, the ideal $\mathfrak{a}$ has a finite resolution by free A-modules; there is an exact sequence

$$0 \to F_r \to F_{r-1} \to \cdots \to F_0 \to \mathfrak{a} \to 0$$

with each F_j a free A-module. Extend this to an exact sequence

$$0 \to (F_r)_f \to \cdots \to (F_0)_f \to \mathfrak{b} \to 0.$$

Since $\mathfrak{b}$ is projective, the sequence splits. We get two free A_f-modules F, F' and an isomorphism $\mathfrak{b} \oplus F \cong F'$. Let r be the rank of F'. Then $A_f \cong \overset{r}{\bigwedge} F' = \mathfrak{b} \otimes_A \overset{r-1}{\bigwedge} F$ by properties of the exterior power $\overset{r}{\bigwedge}$. But $\overset{r-1}{\bigwedge} F \cong A_f$ as well, since rank $F = r-1$. Hence $\mathfrak{b}$ is a free A_f-module of rank 1. Thus $\mathfrak{b}$ is a principal ideal. $\square$

We now prove a result which, when applied to regular integral domains, shows that every divisorial ideal is projective. (Cf. [Bourbaki, 1965, Prop. 1, §3, no. 4].)

Proposition 9.2. *Let A be a Krull domain. If each divisorial ideal is invertible, then $A_\mathfrak{m}$ is factorial for each maximal ideal $\mathfrak{m}$. Conversely, if each $A_\mathfrak{m}$ is factorial, then each finitely generated divisorial ideal is invertible.*

Proof. Let $\mathfrak{b}$ be a divisorial $A_\mathfrak{m}$ ideal. Then $\mathfrak{b} = \mathfrak{a} A_\mathfrak{m}$, where $\mathfrak{a}$ is a divisorial A-ideal (since $\mathrm{Div}\, A \to \mathrm{Div}\, A_\mathfrak{m}$ is a surjection). But $\mathfrak{a}$ is invertible, so $\mathfrak{b}$ is a finitely generated projective module over the local ring $A_\mathfrak{m}$ (local here does not necessarily mean noetherian). Hence $\mathfrak{b}$ is principal. So $A_\mathfrak{m}$ is factorial.

Conversely, let $\mathfrak{a}$ be a finitely generated divisorial A-ideal. Then $\mathfrak{a} A_\mathfrak{m}$ is a divisorial, hence principal, ideal in $A_\mathfrak{m}$. Thus $\mathfrak{a}$ is locally free, so projective. $\square$

Corollary 9.3. *Let A be a Krull domain such that $A_\mathfrak{m}$ is a regular local ring for each maximal ideal $\mathfrak{m}$. Then every finitely generated, divisorial ideal is invertible. In particular, if A is noetherian, then every divisorial ideal is invertible.* $\square$

We now state two results which Kaplansky [1970] uses in order to demonstrate Theorem 9.1 [Kaplansky, 1970, Theorems 178 and 179].

Corollary 9.4. *The integral domain A is factorial if and only if*
(a) *the local rings $A_\mathfrak{m}$ are factorial for all maximal ideals $\mathfrak{m}$,*
(b) *the minimal nonzero prime ideals are finitely generated, and*
(c) *invertible ideals are principal.*

Proof. Use condition (ii) of Theorem 6.1. $\square$

Proposition 9.5. *The integral domain A is factorial if and only if*
(a) *A has the ascending chain condition for principal ideals,*
(b) *all minimal nonzero prime ideals of $A[X]$ are finitely generated,*
(c) *in any localization of $A[X]$, all invertible ideals are principal, and*
(d) *if $\mathfrak{p} \in \operatorname{Spec} A$ admits an $x \in \mathfrak{p}$ such that $\operatorname{Hom}_A(A/\mathfrak{p}, A/xA) \neq 0$,*
then $A_\mathfrak{p}$ is factorial.

Proof. [Kaplansky, 1970, Theorem 179].   ☐

§ 10. Graded Krull Domains and Homogeneous Ideals

In this section we study graded Krull domains, their homogeneous divisorial ideals and the functorial properties of the divisor class group. Applications will be made later. Again, a central tool is Nagata's Theorem.

Let A be a graded commutative ring. Recall that this means $A = \coprod_{n \geq 0} A_n$ where the A_n are abelian groups. Further one has functions $A_n \otimes_{A_0} A_m \to A_{n+m}$ for each pair (m,n) which makes A into a ring. A graded or homogeneous A-module M is an A-module M which admits a decomposition $M = \coprod_n M_n$ such that $A_m M_n \subseteq M_{n+m}$ for each pair (m,n). The elements in $\bigcup_n A_n$ and $\bigcup_n M_n$ are called homogeneous elements.

An ideal $\mathfrak{a}$ in A is homogeneous provided $\mathfrak{a}$ is generated by homogeneous elements or, equivalently, $\mathfrak{a} = \coprod A_n \cap \mathfrak{a}$. A graded homomorphism $f: M \to N$ of two graded A-modules is a homomorphism such that $f(M_n) \subseteq N_{n+d}$ for some fixed integer d and all n.

Lemma 10.1. *Let A be a graded ring. Suppose $\mathfrak{p}$ is a prime ideal. Let $\mathfrak{q}$ be the ideal generated by the homogeneous elements in $\mathfrak{p}$. Then $\mathfrak{q}$ is a prime ideal.*

Proof. Suppose a and b in A are such that $ab \in \mathfrak{q}$. Write $a = \sum a_j$, $b = \sum b_j$ where the a_j and b_j are homogeneous elements. Let i_0, j_0 be the least integers such that $a_{i_0} \notin \mathfrak{q}$ and $b_{j_0} \notin \mathfrak{q}$. Now $(ab)_{i_0 + j_0} = \sum_{r+s=i_0+j_0} a_r b_s$. If $r > i_0$, then $s < j_0$, so $a_r b_s \in \mathfrak{q}$. If $r < i_0$, then $a_r b_s \in \mathfrak{q}$. Hence $a_{i_0} b_{j_0} = (ab)_{i_0+j_0} - \sum_{\substack{r \neq i_0 \\ r+s=i_0+j_0}} a_r b_s$ and this element is in $\mathfrak{q}$. But, since $\mathfrak{p}$ is a prime ideal, one of the homogeneous elements a_{i_0}, b_{j_0} is in $\mathfrak{p}$ and hence in $\mathfrak{q}$. This means that no such pair (i_0, j_0) exists. Hence one of a or b is in $\mathfrak{q}$.   ☐

Proposition 10.2. *Let A be a graded Krull domain. Let $\mathrm{Div_h}(A)$ denote the subgroup of $\mathrm{Div}(A)$ generated by the homogeneous divisorial (prime) ideals and set $\mathrm{Prin_h}(A) = \mathrm{Prin}(A) \cap \mathrm{Div_h}(A)$. Then the inclusion $\mathrm{Div_h}(A) \to \mathrm{Div}(A)$ induces a bijection.*

$$\mathrm{Div_h}(A)/\mathrm{Prin_h}(A) \to \mathrm{Cl}(A).$$

Proof. We can suppose $A \neq A_0$. Let S denote the set of nonzero homogeneous elements. Then $S^{-1}A$ is a graded ring with grading defined by $\mathrm{degree}(a/s) = \mathrm{degree}\,a - \mathrm{degree}\,s$. Let $L = (S^{-1}A)_0$ which is the set $\{a/s: \mathrm{degree}\,a = \mathrm{degree}\,s; \; a \in A, s \in S, a \text{ homogeneous}\}$. In $S^{-1}A$ there are elements of positive degree. Let t be an element of least positive degree in $S^{-1}A$. Then t is transcendental over L. For suppose $l_0 t^r + l_1 t^{r-1} + \cdots + l_r = 0$ with $l_0 \neq 0$, $l_j \in L$ for $0 \leqslant j \leqslant r$. Now $\mathrm{degree}(l_j t^{r-j}) = (r-j) \cdot \mathrm{degree}\,t$. Hence $l_j t^{r-j} \in (S^{-1}A)_{(r-j)\,\mathrm{deg}\,t}$. As $(r-j) \cdot \mathrm{degree}\,t \neq (r-i) \cdot \mathrm{degree}\,t$ when $i \neq j$, it follows that each term $l_i t^{r-i}$ is zero. But then each $l_j = 0$, a clear contradiction.

Hence $(S^{-1}A) \supseteq L[t]$. But $t = a/s$ where a and s are homogeneous and $as \neq 0$. Then $t^{-1} = s/a$ is in $S^{-1}A$. So $(S^{-1}A) \supseteq L[t, t^{-1}]$ as a graded subring. Let us show this is really an equality.

Suppose u in $S^{-1}A$ is homogeneous. Let $m = \mathrm{degree}\,u$. We can find integers q, r with $0 \leqslant r < \mathrm{degree}\,t$ such that $m = q \cdot (\mathrm{degree}\,t) + r$. Then $\mathrm{degree}(u t^{-q}) = \mathrm{degree}\,u - q \cdot (\mathrm{degree}\,t) = r$. It cannot happen that $r > 0$, since we have chosen t to have least positive degree. Hence $r = 0$. Thus $u t^{-q} \in L$. So every homogeneous element in $S^{-1}A$, and hence every element, is in $L[t, t^{-1}]$. So $S^{-1}A$ is a factorial ring.

Nagata's Theorem says that $\mathrm{Cl}(A)$ is generated by those divisorial prime ideals which meet S. Let $\mathfrak{p}$ be a minimal nonzero prime ideal such that $\mathfrak{p} \cap S \neq \emptyset$. Then $\mathfrak{p}$ is a homogeneous ideal, by Lemma 10.1. So $\mathrm{Cl}(A)$ is generated by homogeneous ideals. □

This result has several important consequences. Particularly, it allows one to study the behavior of homogeneous ideals, rather than all ideals.

Corollary 10.3. *Let $A = A_0 + A_1 + \cdots$ be a graded noetherian Krull domain such that A_0 is a field. Let $\mathfrak{m} = A_1 + \cdots$. Then*

$$\mathrm{Cl}(A) \to \mathrm{Cl}(A_\mathfrak{m})$$

is a bijection.

Proof. We know that $\mathrm{Cl}(A) \to \mathrm{Cl}(A_\mathfrak{m})$ is a surjection. We need to know that it is a bijection. According to the proposition we can restrict our attention to homogeneous ideals. Suppose $\mathfrak{a}$ is a homogeneous divisorial ideal which becomes principal in $A_\mathfrak{m}$. Then $\mathfrak{a}$ is free, and hence principal, because of the following lemma. □

Lemma 10.4. *Let $A = A_0 + A_1 + \cdots$ be a noetherian graded ring, where A_0 is a field. Let M be a finitely generated graded A-module. Then M is free if, and only if, $M_\mathfrak{m}$ is projective.*

Proof. If M is free, then $M_\mathfrak{m}$ is free. Suppose, on the other hand, that $M_\mathfrak{m}$ is projective. Let $a_1, \ldots, a_r$ be a subset of the set of homogeneous generators of M which form a basis for $M_\mathfrak{m}$. Then consider the graded homomorphism $A^r \xrightarrow{f} M$ determined by the set $a_1, \ldots, a_r$. Since $f_\mathfrak{m}$ is a bijection, it follows that $(\ker f)$ and $(\operatorname{coker} f)$ are graded A-modules with the property that they are locally zero. We show that they have to be zero.

For let L be a graded A-module of finite type such that $L_n = 0$ for almost all n. Then $L = 0$ if and only if $L/\mathfrak{m}L = 0$. But $L/\mathfrak{m}L = L_\mathfrak{m}/\mathfrak{m}L_\mathfrak{m}$. And $L_\mathfrak{m}/\mathfrak{m}L_\mathfrak{m} = 0$ if and only if $L_\mathfrak{m} = 0$. Hence $L = 0$ if and only if $L_\mathfrak{m} = 0$. □

A result similar to Lemma 10.4 will be needed later.

Lemma 10.4′. *Let A be a graded noetherian ring. A graded finitely generated A-module M is a free A-module if and only if $A_0 \otimes_A M$ is a free A_0-module and $\operatorname{Tor}_1^A(A_0, M) = 0$.* □

Corollary 10.3 has a geometric application (cf. [Samuel, 1964d]). Suppose A is the graded homogeneous coordinate ring of a projective variety. Then $A_\mathfrak{m}$ is the coordinate ring of the projecting cone. Hence A is factorial if and only if the projecting affine cone is factorial.

Corollary 10.5. *Let $A = A_0 + A_1 + \cdots$ be a graded Krull domain such that $A_0 = k$ is a field. Let k' be a field extension of k. Suppose $A \otimes_k k' = A'$ is a Krull domain. Then A' is a faithfully flat A-module and the induced homomorphism*

$$\mathrm{Cl}(A) \to \mathrm{Cl}(A')$$

is an injection.

Proof. The A-module A' is certainly faithfully flat since $M \otimes_A A' = M \otimes_k k'$ for the A-module M. Suppose that $\mathfrak{a}$ is a divisorial A-ideal which is homogeneous and such that $\mathfrak{a}A'$ is principal. Then $\mathfrak{a}$ contains a homogeneous generator for the principal ideal $\mathfrak{a}A'$ by [Cartan and Eilenberg, 1958, Theorem VIII, 6.1]. Hence $\mathfrak{a}$ itself is principal. □

Next we give a theorem due to Samuel [1964a], and then show how this result can be generalized.

Theorem 10.6. *Let* $A = A_0 + A_1 + \cdots$ *be a factorial graded noetherian integral domain. Then*

(a) A_0 *is factorial and*

(b) *Each* A_i *is a divisorial (or reflexive)* A_0-*module.*

Proof (according to Samuel [1964a]).

(a) The ring A_0 is factorial if and only if the intersection of two principal ideals is principal. Let a and b be in A_0 with $ab \neq 0$. Then $aA \cap bA = cA$ for some c in A. But c must be homogeneous and in A_0. Then $cA_0 = aA_0 \cap bA_0$, so A_0 is factorial.

(b) To see that each A_j is divisorial, it is enough to show that each regular A_0-sequence f, g is a regular A_j-sequence. But f, g is a regular A_0-sequence if and only if $fA_0 \cap gA_0 = fgA_0$. Then $fA \cap gA = fgA$. So f, g is a regular A-sequence and hence a regular A_j-sequence for each j. ⬚

We now go on to discuss more general situations. We suppose the graded noetherian integral domain $A = A_0 + A_1 + \cdots + A_n + \cdots$ is normal. Then A_0 is noetherian and normal. We want to know when the extension $A_0 \to A$ satisfies condition (PDE) (§ 6). If $\mathfrak{p} \in X^{(1)}(A)$ and $\mathfrak{p}$ is not homogeneous, then $\mathfrak{p} \cap A_0 = (0)$ by Lemma 10.1. If each A_j is a reflexive A_0-module then (PDE) is satisfied. For suppose $\mathfrak{p}$ is a homogeneous divisorial prime ideal in A. If $\mathrm{ht}(\mathfrak{p} \cap A_0) > 1$, then $\mathfrak{p} \cap A_0$ contains a regular A_0-sequence f, g of length two. But then f, g is a regular A-sequence of length two in A contained in $\mathfrak{p}$, a contradiction to $\mathrm{ht}\,\mathfrak{p} \leqslant 1$. Suppose, on the other hand, that condition (PDE) is satisfied. Then any regular A_0-sequence f, g is a regular A-sequence. For if not, there would be a prime divisorial homogeneous ideal $\mathfrak{p}$ of A such that $\mathfrak{p} \supseteq Af + Ag$. But then $\mathrm{ht}(\mathfrak{p} \cap A_0) \geqslant 2$, contradicting (PDE). But if f, g is a regular A-sequence, then f, g is a regular A_j-sequence for each j.

This discussion is summarized in the following statement.

Proposition 10.7. *Let* $A = A_0 + A_1 + \cdots$ *be a graded noetherian normal integral domain. Then*

(a) A_0 *is a noetherian normal integral domain.*

(b) $A_0 \to A$ *satisfies condition* (PDE) *if and only if each* A_j *is a reflexive* A_0-*module.*

(c) *When* $A_0 \to A$ *satisfies* (PDE) *the induced homomorphism* $\mathrm{Cl}(A_0) \to \mathrm{Cl}(A)$ *is an injection.*

There remains only (c) to verify and that verification depends on a lemma.

Lemma 10.8. *Suppose* $A = A_0 + A_1 + \cdots$ *is a graded noetherian integral domain which is integrally closed. Suppose each* A_i *is a reflexive*

A_0-*module* (it is finitely generated since A is noetherian). *Let* $\mathfrak{a}$ *be an ideal in* A_0. *Then*

$$A_0:(A_0:\mathfrak{a}) \supseteq (A:(A:\mathfrak{a}A)) \cap A_0 \,.$$

In particular, if $\mathfrak{a} = A_0:(A_0:\mathfrak{a})$, *then equality holds.*

Proof. Let k be the field of quotients of A_0. Then $k \otimes_{A_0} A$ is again a graded ring. The module $A:\mathfrak{a}A$ is an A-submodule of $k \otimes_{A_0} A = kA$ and is graded with $(A:\mathfrak{a}A)_n = \{x \in kA_n : x\mathfrak{a} \subseteq A_n\}$. It follows that the graded ideal $A:(A:\mathfrak{a}A)$ has homogeneous components given by $(A:(A:\mathfrak{a}A))_n = \bigcap_{i \geq 0} \{x \in A_n : x(A:\mathfrak{a}A)_i \subseteq A_{n+i}\}$. In particular $(A:(A:\mathfrak{a}A))_0$ $= \bigcap_{i \geq 0} \{x \in A_0 : x(A:\mathfrak{a}A)_i \subseteq A_i\}$. Hence $A_0:(A_0:\mathfrak{a}) \supseteq (A:(A:\mathfrak{a}A))_0 \supseteq \mathfrak{a}$. $\square$

Corollary 10.9. (a) *If* $\mathfrak{p} \in X^{(1)}(A_0)$, *there is a* $\mathfrak{q}$ *in* $X^{(1)}(A)$ *such that* $\mathfrak{q} \cap A_0 = \mathfrak{p}$. *Thus* $\mathrm{Div}(A_0) \to \mathrm{Div}(A)$ *is a monomorphism.*

(b) *If* $\mathfrak{a}$ *is a divisorial ideal such that* $A:(A:A\mathfrak{a})$ *is principal, then* $\mathfrak{a}$ *is principal.*

Proof. (a) Since $A:(A:A\mathfrak{p}) \cap A_0 = \mathfrak{p}$ the ideal $A:(A:A\mathfrak{p})$ is a proper divisorial ideal in A. Hence it is contained in a $\mathfrak{q}$ in $X^{(1)}(A)$, which is necessarily homogeneous. But then $\mathfrak{q} \cap A_0 = \mathfrak{p}$.

(b) If $A:(A:\mathfrak{a}A)$ is principal, a generator must lie in A_0. But then $\mathfrak{a}$ is principal. $\square$

Proof of (c) *of Proposition* 10.7. We have the following commutative diagram with exact rows and columns

$$
\begin{array}{ccccccccc}
 & & 0 & & 0 & & & & \\
 & & \downarrow & & \downarrow & & & & \\
0 & \to & \mathrm{Prin}\,A_0 & \to & \mathrm{Div}\,A_0 & \to & \mathrm{Cl}(A_0) & \to & 0 \\
 & & \downarrow & & \downarrow & & \downarrow & & \\
0 & \to & \mathrm{Prin}_h A & \to & \mathrm{Div}_h A & \to & \mathrm{Cl}(A) & \to & 0 \,.
\end{array}
$$

Let $\mathfrak{a}$ be a divisorial ideal whose class becomes 0 in $\mathrm{Cl}(A)$. That is to say, $A:(A:\mathfrak{a}A)$ is principal. But then $\mathfrak{a}$ itself is principal. Hence $\mathrm{Cl}(A_0) \to \mathrm{Cl}(A)$ is a monomorphism. $\square$

Corollary 10.10. *If* A *is factorial, then so is* A_0 (see Proposition 10.6). $\square$

We now consider a noetherian normal integral domain A and a finitely generated A-module M. Let $S_A(M)$ denote the symmetric algebra. Then $S_A(M)$ is a graded noetherian A-algebra with $S^0(M) = A, S^1(M) = M$ and $S^n(M)$ the nth symmetric power. We will now prove a generalization of Samuel's Théorème 2 in [1964a].

Theorem 10.11. *Let A be a noetherian Krull domain and M a finitely generated module. Suppose each $S^n(M)$ is a reflexive A-module. Then $S_A(M)$ is a Krull domain* (noetherian) *and $\mathrm{Cl}(A) \to \mathrm{Cl}(S_A(M))$ is a bijection.*

Proof. Since each component of $S_A(M)$ is torsion free $S_A(M) \subseteq U^{-1}S_A(M) = S(U^{-1}M)$ where $U = A - \{0\}$. But $S(U^{-1}M)$ is a ring of polynomials over the field of quotients of A. Hence $S_A(M)$ is an integral domain. Since $S^n_A(M)_\mathfrak{p} = S^n_{A_\mathfrak{p}}(M_\mathfrak{p})$ for each $\mathfrak{p}$ and since $S^n(M) = \bigcap_{\mathfrak{p} \in X^{(1)}(A)} S^n_{A_\mathfrak{p}}(M_\mathfrak{p})$, we get $S_A(M) = \bigcap_{\mathfrak{p} \in X^{(1)}(A)} S_{A_\mathfrak{p}}(M_\mathfrak{p})$. But each $S_{A_\mathfrak{p}}(M_\mathfrak{p})$ is a ring of polynomials over the principal valuation ring $A_\mathfrak{p}$. Hence $S_A(M)$ is integrally closed.

We have seen that $\mathrm{Cl}(A) \to \mathrm{Cl}(S_A(M))$ is a monomorphism. We will show now that it is a surjection. It is sufficient to prove that $(\mathfrak{p}S_A(M))^{**}$, the divisorial ideal associated to $\mathfrak{p}$ in $X^{(1)}(A)$, is prime. We need another lemma.

Lemma 10.12. *Assume the same notation and hypotheses as before. Let $\mathfrak{a}$ be a homogeneous divisorial ideal in $S_A(M)$. Then $\mathfrak{a} = \bigcap_{\mathfrak{p} \in X^{(1)}(A)} \mathfrak{a}_\mathfrak{p}$.*

Proof. Suppose $\mathfrak{a} \neq \bigcap \mathfrak{a}_\mathfrak{p}$. Choose x in $\bigcap_{\mathfrak{p} \in X^{(1)}(A)} \mathfrak{a}_\mathfrak{p}$ with $x \notin \mathfrak{a}$, x homogeneous, and such that $\mathfrak{P} = \{y \in S_A(M) : yx \in \mathfrak{a}\}$ is maximal among the ideals sending an element not in $\mathfrak{a}$ into $\mathfrak{a}$. Then $\mathfrak{P}$ is a homogeneous prime divisorial ideal. (It is divisorial since $\mathfrak{a}$ is divisorial.) Suppose f, g in $S(M)$ are such that $fg \in \mathfrak{P}$. Suppose that $g \notin \mathfrak{P}$. Then $fgx \in \mathfrak{a}$ but $gx \notin \mathfrak{a}$. Now $(\mathfrak{P} + fS(M))gx \subseteq \mathfrak{a}$. But then $f \in \mathfrak{P}$. Now $\mathfrak{P}_\mathfrak{p} = S(M)_\mathfrak{p}$ for all $\mathfrak{p}$ in $X^{(1)}(A)$. Hence $\mathfrak{P} \cap A = (0)$. But $\mathrm{ht}\,\mathfrak{P} \leqslant 1$ and $\mathfrak{P}$ is homogeneous. So $\mathfrak{P} \cap A \neq (0)$, a contradiction unless $\mathfrak{P} = S_A(M)$. In that case $x \in \mathfrak{a}$. □

Return to the proof of the theorem. Let u, v be homogeneous elements such that $uv \in (\mathfrak{p}S(M))^{**}$. Suppose $v \notin (\mathfrak{p}S(M))^{**}$. Let $\mathfrak{q}$ be in $X^{(1)}(A)$. Then $(\mathfrak{p}S(M))^{**}_\mathfrak{q} = \mathfrak{p}S(M_\mathfrak{q})$. This is $S(M_\mathfrak{q})$ if $\mathfrak{q} \neq \mathfrak{p}$. Hence $uv \in \mathfrak{p}S(M_\mathfrak{p})$. As $v \in S(M_\mathfrak{q})$ for all $\mathfrak{q} \neq \mathfrak{p}$, it cannot happen that $v \in \mathfrak{p}S(M_\mathfrak{p})$. For otherwise $v \in (\mathfrak{p}S(M))^{**}$. But $\mathfrak{p}S(M_\mathfrak{p})$ is a prime ideal in $S(M_\mathfrak{p})$. Hence $u \in \mathfrak{p}S(M_\mathfrak{p})$ and, as $u \in S(M_\mathfrak{q})$ for $\mathfrak{q} \neq \mathfrak{p}$, we get $u \in (\mathfrak{p}S(M))^{**}$ by applying the lemma.

This implies that $\mathrm{Div}(A) \to \mathrm{Div}_\mathrm{h}S(M)$ is an isomorphism. Hence $\mathrm{Cl}(A) \to \mathrm{Cl}(S_A(M))$ is an isomorphism. □

Samuel [1964a] actually demonstrated that $S(M)$ is a Krull domain when each $S^n(M)$ is reflexive. He also has an example which shows that (a) $S(M)$ can be an integrally closed noetherian integral domain but not every $S^n(M)$ is reflexive, and therefore (b) $A \to S(M)$ does not satisfy (PDE) and for which (c) $\mathrm{Cl}(A) = 0$ and $\mathrm{Cl}(S(M)) = \mathbb{Z}$. In fact, let

A be a regular local ring of dimension 2. Let f, g generate the maximal ideal of A. Let M be the module which is generated by x, y with $fx + gy = 0$. Then $S(M) = A[X, Y]/(aX + bY)$. As is seen in Section 14 or as can be seen from [J. Sally, 1971] $S(M)$ is a normal ring. The ideal in $S(M)$ generated by f and g is prime, of height one, and it generates $\mathrm{Cl}(S(M))$, which is $\mathbb{Z}$. But $\mathfrak{P} \cap A = (f, g)$. So $\mathrm{ht}(\mathfrak{P} \cap A) = 2$. Thus $A \to S(M)$ does not satisfy condition (PDE). Now $S^1(M)$ is not free, and hence not reflexive.

We use the results in this section to verify an exercise in Bourbaki [1965, p. 99, ex. 7].

Let A be a Krull domain. Let $w_1, \ldots, w_n$ be weights assigned to $X_1, \ldots, X_n$ respectively in $A[X_1, \ldots, X_n]$. Suppose $F(X_1, \ldots, X_n)$ in $A[X_1, \ldots, X_n]$ is a weighted homogeneous prime element of weight w. That F is weighted homogeneous with weight w is equivalent to the condition $F(Z^{w_1} X_1, \ldots, Z^{w_n} X_n) = Z^w F(X_1, \ldots, X_n)$ in $A[X_1, \ldots, X_n, Z]$.

Proposition 10.13. *Let A be a Krull domain and F a weighted homogeneous prime element in $A[X_1, \ldots, X_n]$. Suppose p is a positive integer prime to the weight w of F. Suppose further that either $p \equiv 1$ (mod weight F) or projective A-modules are free. Then $A[X_1, \ldots, X_n, Z]/(Z^p - F)$ is a Krull domain whose divisor class group is isomorphic to $\mathrm{Cl}(A)$ under the canonical homomorphism.*

Proof. Suppose w is the weight of F. Let B denote $A[x_1, \ldots, x_n, z]$ with $z^p = F(x_1, \ldots, x_n)$. Then note that $B/Bz = A[X_1, \ldots, X_n]/(F)$. Hence z is a prime element in B. Thus it is sufficient to show that $B[z^{-1}]$ is a Krull domain with $\mathrm{Cl}(B[z^{-1}]) = \mathrm{Cl}(A)$, according to Nagata's Theorem 7.1.

Suppose first that $p \equiv 1$ (mod w) say $p = 1 + rw$ for a suitable r. Then $z^p = z^{1+rw}$ or $z = z^{-rw} z^p$. But as $z^p = F(x_1, \ldots, x_n)$ and as F is homogeneous of weight w, we get $z = F(x_1 z^{-rw_1}, \ldots, z_n z^{-rw_n})$. For each i, let $y_i = x_i z^{-rw_i}$. Then $A[x_1, \ldots, x_n, z, z^{-1}] = A[y_1, \ldots, y_n, F(y_1, \ldots, y_n)^{-1}]$. Thus $B[z^{-1}]$ is the localization at the prime element $F(y_1, \ldots, y_n)$ of the pure transcendental extension $A[y_1, \ldots, y_n]$. Hence $\mathrm{Cl}(A) \to \mathrm{Cl}(B[z^{-1}])$ is a bijection.

Then suppose that projective A-modules are free. Since p is relatively prime to w, there is a positive integer p' such that $p' p \equiv 1$ (mod w). Let $B' = B[T]/(T^{p'} - z)$. Since z is a prime element in B, the extension B' is an integral domain which is free and finitely generated as a B-module. Furthermore, $B' \cong A[x_1, \ldots, x_n, t]$ with $t^{pp'} = F(x_1, \ldots, x_n)$. By the above calculation, B' has divisor class group $\mathrm{Cl}(A)$. The extension $B \to B'$ is faithfully flat.

After redefining the weights of the x_i to be $pp' w_i$, assigning the weight $p'w$ to z and the weight w to t, we get that both B and B' are graded Krull domains and that the extension $B \to B'$ is homogeneous. By Proposition 10.2 the divisor class groups $\mathrm{Cl}(B)$ and $\mathrm{Cl}(B')$ are generated by classes of homogeneous ideals. Thus an element in the kernel of $\mathrm{Cl}(B) \to \mathrm{Cl}(B')$ can be represented by the class of a homogeneous ideal $\mathfrak{a}$ which, by faithfully flat descent, must be finitely generated and invertible. Hence $A \otimes_B \mathfrak{a}$ is free and finitely generated by the hypothesis on A. As in the proof of Lemma 10.4, we can find a graded homomorphism $f: B^r \to \mathfrak{a}$ whose cokernel C has the property that $A \otimes_B C = 0$ and then $C = 0$. Hence f is a surjection onto a projective. Therefore the kernel of f satisfies the same properties as does the cokernel and is thus zero. So f is a bijection. Hence $\mathfrak{a}$ is a principal ideal and therefore $\mathrm{Cl}(B) \to \mathrm{Cl}(B')$ is an injection. From the first part of the proof it is seen that the composition $\mathrm{Cl}(A) \to \mathrm{Cl}(B) \to \mathrm{Cl}(B')$ is a bijection and it has just been seen that the second homomorphism is an injection. Therefore both maps are bijections. $\square$

Corollary 10.14. *Suppose r, s, t are positive integers which are pairwise relatively prime. Let A be a Krull domain. Then $A[x, y, z]$ with $x^r + y^s + z^t = 0$ is a Krull domain and $\mathrm{Cl}(A) \to \mathrm{Cl}(A[x, y, z])$ is a bijection in case projective A-modules are free or one of $t \equiv 1 \pmod{rs}$, $s \equiv 1 \pmod{rt}$, $r \equiv 1 \pmod{st}$ is satisfied. In particular, if A is a regular local ring then $A[x, y, z]$ is factorial.*

Proof. It is sufficient to show that $X^r + Y^s$ is a prime element in $K[X, Y]$, where K is the field of quotients of A. This is equivalent to showing that $X^r + Y^s$ is irreducible and this follows from the next lemma. $\square$

Lemma 10.15. *Let A be an integral domain. Suppose r and s are relatively prime positive integers. Then $X^r + Y^s$ is irreducible in $A[X, Y]$.*

Proof. Attach the weight s to X and the weight r to Y. Then $B = A[X, Y]$ becomes a graded ring with B_n generated as a free A-module by the monomials $X^i Y^j$ with $is + jr = n$, i and j being non-negative integers. The element $X^r + Y^s$ is a (weighted) homogeneous element of degree rs. Suppose it has a factorization $X^r + Y^s = fg$. Then both f and g are homogeneous, say of degree n and m respectively. Then $n + m = rs$. Suppose $f = \sum_{(i,j)} a_{i,j} X^i Y^j$ with $is + jr = n$. There must be a term not involving Y (and one not involving X), say $a_p X^p$ with $ps = n$ and $a_p \neq 0$. Then we can write $f = \sum_{i=0}^{p} a_i Y^{q_i} X^i$ with $is + q_i r = ps = n$. Since r and s are relatively prime, the integer s must divide q_i and r must divide

$p-i$. But p and i both are less than r. Hence $p=i$ and $q_i=0$. But then $f=a_p X^p$ and then X must divide Y also, a contradiction. □

The examples of factorial rings constructed in Corollary 10.14 will be used in Chapter IV to show the existence of factorial rings for which $Cl(A) \to Cl(A[[T]])$ is not an isomorphism. Rings of this type also arise in the sections on Galois and radical descent. There the class groups of these rings are calculated in cases where the integers are not relatively prime.

§ 11. Quadratic Forms

Let k be a field with char $k \neq 2$. Let F be a non-degenerate quadratic form in $k[X_1, ..., X_n]$. We want to study the ring $A_F = k[X_1, ..., X_n]/(F)$. Thus we can assume, by a linear change of variables if necessary, that $F = X_1^2 + b X_2^2 + \cdots$. We then show that A_F is normal in case $n \geqslant 3$. Then we study the class group $Cl(A_F)$. It is seen that $Cl(A_F) = (0)$ for $n \geqslant 5$. (Klein-Nagata. See [Samuel 1964d].) When $n=2$ and A_F is an integral domain, then its integral closure is a principal ideal domain. We will then study the cases $n=3$ and $n=4$. In both cases $Cl(A_F)$ is cyclic. The exact determination depends on the arithmetic of the field. Thus, for a real closed field a complete determination can be made.

The first result is a technical lemma which will be used to show that A_F is normal. It is relatively well known but does not seem to appear in the literature.

Lemma 11.1. *Let A be a factorial ring in which 2 is a unit, a in A a square free element (i.e., if p is a prime element in A, then $a \notin p^2 A$) which is not a unit. Then $A[T]/(T^2 - a)$ is normal.*

Proof. The demonstration consists of showing that any element in the field of quotients of $A[T]/(T^2 - a)$ which is integral is in $A[T]/(T^2 - a)$.

Let L be the field of quotients of $A[T]/(T^2 - a)$. We denote by t the image of T in $A[T]/(T^2 - a)$ (and in L). Let $B = A[t] = A[T]/(T^2 - a)$. Then $L = K[t]$. Any element in L can be written uniquely as a sum $r + st$, r, s in K. Suppose $r + st$ is integral over A. Then the minimal polynomial of $r + st$ has its coefficients in A. Now the minimal polynomial of $r + st$ is $X^2 - 2rX + (r^2 - as^2)$, provided $s \neq 0$. Hence $r + st$ is integral over A if and only if $-2r \in A$ and $r^2 - as^2 \in A$. Now $r \in A$ immediately. To see that $s \in A$, we show that whenever $s = s_1/s_2$ with s_1 and s_2 relatively prime elements of A, then s_2 is a unit. For $-as^2 \in A$, say $+as^2 = h$. Then $a s_1^2 = h s_2^2$. Suppose the prime element p divides s_2. Then $a s_1^2 \in p^2 A$. Since s_1 and s_2 are relatively prime, $a \in p^2 A$. But this is a contradiction.

Hence s_2 has no prime factors, so it is a unit. Thus $s \in A$. Hence the integral closure of A in L is $A[t]$. □

Remark. When 2 is not a unit in A this argument becomes more difficult. Then one needs to know the factorization of 2 into prime elements and the properties of the polynomial $T^2 - a$ in $(A/2A)[T]$.

We use this lemma to show that A_F is normal when $n \geqslant 3$.

Proposition 11.2. *Let F be a non-degenerate quadratic form in $k[X_1, \ldots, X_n]$. Assume that $n \geqslant 3$. Then $k[X_1, \ldots, X_n]/(F)$ is normal.*

Proof. We can assume $F = X_1^2 + F_1$ where F_1 is a non-degenerate quadratic form in $k[X_2, \ldots, X_n]$. Since $n \geqslant 3$, the form F_1 cannot be a square of a linear polynomial. Hence F_1 is square free. The normality follows from the lemma. □

The case $n = 2$ is quite different. Then we assume $F = X_1^2 - a X_2^2$. Now F need not be a prime element in $k[X_1, X_2]$. If F is not prime, then $F = UV$ where $U = X_1 + r X_2, V = X_1 - r X_2, r^2 = a$. Then $k[X_1, X_2]/(F) = k[U, V]/(UV)$. If, on the other hand, F is irreducible (i.e., $T^2 - a$ is a prime in $k[T]$) then $k[X_1, X_2]/(F)$ is at least an integral domain. Let x_1 and x_2 be the images in A_F of X_1 and X_2 respectively. Then, in the field of quotients, L, of A_F, we have the element $x_1/x_2 = t$ whose square, $(x_1/x_2)^2$, is a. Thus $L \supseteq k(\sqrt{a}) = k(t)$, a quadratic extension of k. Now t is integral over A_F, for $t^2 = a \in k \subseteq A_F$. Hence the integral closure of A_F contains $k(t)$. The element x_1 is transcendental over $k \subset L$, and hence over $k(t)$. Hence $k(t)[x_1]$ is integral over A_F. Now $x_2 = x_1/t$. So $k(t)[x_1] \supseteq A_F = k[x_1, x_2]$. As $k(t)[x_1]$ is a principal ideal domain, it is integrally closed. As it is integral over A_F, it must be the integral closure of A_F in L.

We now state and prove the Klein-Nagata result.

Proposition 11.3. *Let F be a non-degenerate quadratic form in $k[X_1, \ldots, X_n]$. If $n \geqslant 5$, then $A_F = k[X_1, \ldots, X_n]/(F)$ is a factorial ring.*

Proof. The ring A_F is a graded Krull domain with $(A_F)_0 = k$, a field. Let $\bar{k}$ be the algebraic closure of k. Then $\mathrm{Cl}(A_F) \to \mathrm{Cl}(A_F \otimes_k \bar{k})$ is an injection. Thus we may assume that $k = \bar{k}$, and show that A_F is factorial in this case.

Since k is algebraically closed, we can write $F = UV + G$ where $UV = X_1^2 + b X_2^2$ and $G \in k[X_3, X_4, X_5, \ldots, X_n]$ and is a prime element. Then $A_F = k[U, V, X_3, \ldots, X_n]/(UV + G)$. Let $u, v, x_3, \ldots, x_n$ denote the images of $U, V, X_3, \ldots, X_n$ in A_F. Now $A_F/u A_F \cong k[V][X_3, \ldots, X_n]/(G)$ is an integral domain (even normal). Hence u is a prime element in A_F.

By Nagata's Theorem, Corollary 7.2, $Cl(A_F) \cong Cl(A_F[u^{-1}])$. But $A_F[u^{-1}] = k[u, v, x_3, \ldots, x_n][u^{-1}]$ where $uv + G(x_3, \ldots, x_n) = 0$. Hence $A_F[u^{-1}] = k[x_3, \ldots, x_n, u, u^{-1}]$. Since $u, x_3, \ldots, x_n$ are algebraically independent, then $A_F[u^{-1}]$ is a factorial ring. Hence A_F is factorial. $\square$

The rings A_F are not necessarily factorial for $n = 3$ and $n = 4$. Suppose $n = 4$, that $k = \bar{k}$ and that $F = X_1^2 + X_2^2 + X_3^2 + X_4^2$. Then we can write $X_1^2 + X_2^2 = XU$, $X_3^2 + X_4^2 = YV$ where $X_1 + iX_2 = X$, $X_1 - iX_2 = U$, $X_3 + iX_4 = Y$, $X_3 - iX_4 = V$ and $i^2 + 1 = 0$. It is seen, in Section 14, where this ring is studied, that $Cl(A_F) \cong \mathbb{Z}$.

Let us consider here, the ring $A_F = k[X_1, X_2, X_3]/(F)$ where $F = X_1^2 + X_2^2 + X_3^2$ and $k = \bar{k}$. Set $X = X_1 + iX_2$, $V = X_1 - iX_2$, as before. Then $A_F = k[x, v, x_3]$ with $xv + x_3^2 = 0$. Now $A_F[v^{-1}] = k[x_3, v, v^{-1}]$, where x_3, v are algebraically independent. From Nagata's Theorem, we see that $Cl(A_F)$ is generated by the classes of divisorial prime ideals which contain v. But $k[x, v, x_3]/(v) = k[T, X_3]/(X_3^2)$. We want a minimal prime ideal in this ring. But such a minimal prime ideal is generated by the image of X_3. Hence the only prime divisorial ideal in A_F which contains v is the ideal $(v, x_3) = \mathfrak{p}$. Now $\mathfrak{p}^2 = (v^2, vx_3, x_3^2) = v(v, x_3, x)$. Hence $\mathfrak{p}^{(2)} = (v)$, a principal ideal. Consequently $Cl(A_F)$ is generated by the divisor of $\mathfrak{p}$, which satisfies $2 \operatorname{div}(\mathfrak{p}) = 0$. We can conclude that $Cl(A_F) = \mathbb{Z}/2\mathbb{Z}$ provided that we can show that $\mathfrak{p}$ itself is not principal. But suppose $(v, x_3) = (f)$. The element f must be homogeneous. As degree $v = \deg x_3 = 1$, f must be homogeneous of degree 1. Then, since $v = rf$ and $x_3 = sf$ for r, s homogeneous of degree 0, we have r and s are units. Hence $v = vs^{-1}x_3$ where $rs^{-1} \in k$. Then, since $xv + x_3^2 = 0$, we have $x = -sr^{-1}x_3$. Thus $(x, v, x_3) = (x_3)$. But (x, v, x_3) is a maximal ideal. Hence x_3 is a prime element in A_F. This is a contradiction, since $A_F/x_3 A_F = k[X, V]/(XV)$.

The conclusion one draws from these results is that $Cl(A_F)$ is cyclic.

Proposition 11.4. *Let k be a field of characteristic different from 2.*

(a) *If F is a non-degenerate quadratic form in $k[X_1, X_2, X_3]$, then $Cl(A_F)$ is cyclic of order 2 or else A_F is factorial. If k is algebraically closed then $Cl(A_F) = \mathbb{Z}/2\mathbb{Z}$.*

(b) *If F is a non-degenerate quadratic form in $k[X_1, X_2, X_3, X_4]$, then $Cl(A_F)$ is either infinite cyclic or is zero. It is infinite cyclic if k is algebraically closed.* $\square$

Samuel [1964d] uses geometric arguments to analyse the plane conics or the quadratic forms. It is possible to carry out a proof, along the same lines as in the above paragraph, to demonstrate the result of Samuel. The details are not sufficiently interesting to warrant repeating them here. Let us state Samuel's result.

Proposition 11.5 (Samuel). *Let F be a non-degenerate quadratic form in $k[X_1, X_2, X_3]$. Let $A_F = k[X_1, X_2, X_3]/(F)$. Then $\mathrm{Cl}(A_F) = \mathbf{Z}/2\mathbf{Z}$ if and only if there is a nontrivial solution (a rational point) to $F(X_1, X_2, X_3) = 0$ in k. If no such solution exists then A_F is factorial.* ☐

A corresponding theorem does not hold for quadratic forms of dimension 4, as is seen in the next result.

Proposition 11.6. *Let k be a real closed field, F a non-degenerate quadratic form in $k[X_1, X_2, X_3, X_4]$. Let S_F denote the signature of F.*
(a) *If $S_F = 4$ or 0, then $\mathrm{Cl}(A_F) = \mathbf{Z}$.*
(b) *If $S_F = 2$, then A_F is factorial.*

Remark. If $S_F = 4$, then F has no nontrivial rational zero. If $S_F = 2$, then F has nontrivial rational zeros. Likewise F has rational zeros if $S_F = 0$.

Proof. If $S_F = 2$, then we can suppose $F = X_1^2 + X_2^2 + X_3^2 - X_4^2$. Let $U = X_3 + X_4$, $V = X_3 - X_4$. Then $A_F \cong k[X_1, X_2, U, V]/(X_1^2 + X_2^2 + UV)$. As in the proof of the Klein-Nagata result, the image of U is a prime element in A_F, and $A_F[u^{-1}]$ is factorial.

Suppose $S_F = 0$. Then we can suppose $F = X_1^2 - X_2^2 + X_3^2 - X_4^2 = XY + UV$ where $X = X_1 + X_2$, $Y = X_1 - X_2$, $U = X_3 + X_4$, $V = X_3 - X_4$. Then $\mathrm{Cl}(A_F) = \mathbf{Z}$ (see Section 14).

In the case $S_F = 4$, we use a proof, due to L. Roberts, to show that the class group is infinite. Let $\bar{k}$ be the algebraic closure of k. Let $\bar{A}_F = \bar{k} \otimes_k A_F$. Then $\mathrm{Cl}(\bar{A}_F) = \mathbf{Z}$, and the generator is $\mathrm{cl}(\mathfrak{P})$ where $\mathfrak{P} = (x_1 + ix_2, x_3 + ix_4)$. Let $\mathfrak{P}' = (x_1 - ix_2, x_3 - ix_4)$ and $\mathfrak{P}_1 = (x_1 - ix_2, x_3 + ix_4)$. Then $\mathfrak{P} \cap \mathfrak{P}_1 = (x_3 + ix_4)$ and $\mathfrak{P}' \cap \mathfrak{P}_1 = (x_1 - ix_2)$. Hence $\mathrm{cl}(\mathfrak{P}) = \mathrm{cl}(\mathfrak{P}')$. Consider the ideal $\mathfrak{P} \cap \mathfrak{P}'$. This ideal contains the ideal generated by the elements $x_1^2 + x_2^2$, $x_3^2 + x_4^2$, $x_1 x_3 + x_2 x_4$, $x_1 x_4 - x_2 x_3$ which is $\mathfrak{P} \cdot \mathfrak{P}'$. Let $\mathfrak{p}$ denote the ideal in A_F generated by these elements. It can be seen that $\mathfrak{p}$ is a prime ideal in A_F. Hence it is a divisorial ideal. Thus $\mathfrak{p} \bar{A}_F = \mathfrak{P} \cap \mathfrak{P}'$ is divisorial, so $\mathrm{cl}(\mathfrak{p} \bar{A}_F) = 2\,\mathrm{cl}(\mathfrak{P})$. This shows that $\mathrm{Cl}(A_F) \to \mathrm{Cl}(\bar{A}_F)$ is not the zero map. Hence $\mathrm{Cl}(A_F) \cong \mathbf{Z}$. ☐

Corollary 11.7. *If k is a real field and F is a quadratic form of dimension 4 with signature two, then $k[X_1, X_2, X_3, X_4]/(F)$ is factorial.* ☐

The situation for a general field still seems unclear. For example, the form $X_1^2 + X_2^2 + X_3^2 + 2X_4^2$ over $\mathbf{Q}$ yields a factorial ring $A_F = \mathbf{Q}[X_1, X_2, X_3, X_4]/(X_1^2 + X_2^2 + X_3^2 + 2X_4^2)$. To see this, we imbed $\mathbf{Q}$ in the (3)-adic completion k of $\mathbf{Q}$. Over this field, -2 is a square but $X_1^2 + X_2^2$ is still a prime. Hence, as in the signature equal to two case, $k \otimes_\mathbf{Q} A_F$ is factorial. So A_F is factorial.

On the other hand $X_1^2 + X_2^2 + 2(X_3^2 + X_4^2)$ over $\mathbb{Q}$ yields the ring $\mathbb{Q}[X_1, X_2, X_3, X_4]/(X_1^2 + X_2^2 + 2X_3^2 + 2X_4^2)$ with divisor class group $\mathbb{Z}$ (generated by the same ideal $(x_1^2 + x_2^2, x_1 x_3 + x_2 x_4, x_1 x_4 - x_2 x_3, x_3^2 + x_4^2))$.

Problem. Let k be a field, $F(X_1, X_2, X_3, X_4)$ a non-degenerate quadratic form. Find necessary and sufficient conditions in order that $k[X_1, X_2, X_3, X_4]/(F)$ is a factorial ring.

As an application of these results, let us study the rings $A_n = \mathbb{R}[X_0, X_1, \ldots, X_n]/(X_0^2 + \cdots + X_n^2 - 1)$ and $B_n = \mathbb{C}[X_0, \ldots, X_n]/(X_0^2 + \cdots + X_n^2 - 1)$. The ring A_n can be thought of as the ring of polynomial functions on the sphere $S^n = \{(a_0, a_1, \ldots, a_n): a_j \in \mathbb{R}, \sum a_i^2 = 1\}$.

Proposition 11.8. (a) *If* $n \geqslant 3$, *then* $\mathbb{R}[X_0, \ldots, X_n]/(X_0^2 + \cdots + X_n^2 - 1)$ *and* $\mathbb{C}[X_0, \ldots, X_n]/(X_0^2 + \cdots + X_n^2 - 1)$ *are factorial.*
(b) $\mathrm{Cl}(\mathbb{R}[X_0, X_1]/(X_0^2 + X_1^2 - 1)) \cong \mathbb{Z}/2\mathbb{Z}$ *while* $\mathbb{C}[X_0, X_1]/(X_0^2 + X_1^2 - 1)$ *is factorial.*
(c) $\mathbb{R}[X_0, X_1, X_2]/(X_0^2 + X_1^2 + X_2^2 - 1)$ *is factorial while* $\mathrm{Cl}(\mathbb{C}[X_0, X_1, X_2]/(X_0^2 + X_1^2 + X_2^2 - 1)) \cong \mathbb{Z}$.

Proof. Let C_n be either A_n or B_n. Then $\mathrm{Cl}(C_n) = \mathrm{Cl}(C_n[t, t^{-1}])$ where t is an indeterminate. Now $C_n[t, t^{-1}] = k[y_0, y_1, \ldots, y_n, t, t^{-1}]$, where $k = \mathbb{C}$ or $\mathbb{R}$, $y_j = x_j t$ for $j = 0, 1, \ldots, n$ and where $y_0^2 + y_1^2 + \cdots + y_n^2 - t^2 = 0$. If the number of variables, $n + 2$, is greater than or equal to 5, then $C_n[t, t^{-1}]$ is factorial by the Klein-Nagata result, Proposition 11.3. Hence (a) is true.

Now $C_2[t, t^{-1}] = k[y_0, y_1, y_2, t, t^{-1}]$; $y_0^2 + y_1^2 + y_2^2 - t^2 = 0$. Since t is a prime element in $k[y_0, y_1, y_2, t]$; $y_0^2 + y_1^2 + y_2^2 - t^2 = 0$, the group $\mathrm{Cl}(C_2[t, t^{-1}]) \cong \mathrm{Cl}(k[y_0, y_1, y_2, t])$.

By Proposition 11.4, this group is $\mathbb{Z}$ if k is algebraically closed, i.e., if $k = \mathbb{C}$. By Proposition 11.6, $\mathbb{R}[y_0, y_1, y_2, t]$; $y_0^2 + y_1^2 + y_2^2 = t^2$ is factorial. This proves (b).

Now consider $\mathbb{C}[X_0, X_1]/(X_0^2 + X_1^2 - 1)$. Let $u = x_0 + i x_1, v = x_0 - i x_1$. Then $\mathbb{C}[x_0, x_1] = \mathbb{C}[u, v]$; $uv = 1$. Hence $\mathbb{C}[x_0, x_1] = \mathbb{C}[u, u^{-1}]$, which is factorial. However $\mathrm{Cl}(\mathbb{R}[X_0, X_1]/(X_0^2 + X_1^2 - 1)) = \mathrm{Cl}(\mathbb{R}[y_0, y_1, t, t^{-1}]$; $y_0^2 + y_1^2 = t^2) \cong \mathrm{Cl}(\mathbb{R}[y_0, y_1, t])$, this last since t is a prime element. By Proposition 11.5 this class group is $\mathbb{Z}/2\mathbb{Z}$. $\quad \square$

§ 12. Murthy's Theorem

This section is devoted to the demonstration of a remarkable result proved by P. V. Murthy [1964]. It states that a factorial ring which is a factor of a regular ring and which is Cohen-Macaulay is Gorenstein.

For whatever it's worth, a slight improvement is offered here (with Murthy's proof).

The first proposition is a characterization of Cohen-Macaulay rings which are factor rings of Gorenstein rings. It is classical but there seems to be no classical reference. The proof here is a modification of ideas obtained from [Grothendieck, 1967].

Recall that a local noetherian ring B is *Gorenstein* if the injective dimension of B, $\mathrm{id}_B B$, is finite. Then $\mathrm{id}_B B = \mathrm{depth}\, B$ and B is Cohen-Macaulay [Bass, 1963]. Let $0 \to B \to E^0 \to E^1 \to \cdots \to E^n \to 0$ be an injective resolution of B and denote by $E^\bullet$ the complex obtained by deleting B. Let A be a B-algebra. The complex $K_A = \mathrm{Hom}_B(A, E^\bullet)$ is a complex of injective A-modules whose homology is $\mathrm{Ext}_B^\bullet(A, B)$.

Suppose A is a factor ring of B. Let d be dim A. The module of dualizing differentials is the A-module $\mathrm{Ext}_B^{n-d}(A, B)$ and is denoted by $\Omega(A)$.

Proposition 12.1. *The factor ring A is Cohen-Macaulay if and only if the complex K_A is an injective resolution of $\Omega(A)$. When this is the case, then the endomorphism ring $\mathrm{Hom}_A(\Omega(A), \Omega(A)) \cong A$. Furthermore A is Gorenstein if and only if $A \cong \Omega(A)$.*

Proof. What must be shown is that A is Cohen-Macaulay if and only if $\mathrm{Ext}_B^j(A, B) = 0$ for $j \neq n - d$.

Suppose, in fact, that the complex K_A is an injective resolution of $\Omega(A)$. Then $\mathrm{id}_A \Omega(A) = \mathrm{depth}\, A$ [Bass, 1963]. But $\mathrm{id}_A \Omega(A) = n - (n-d) = d$. Hence dim $A = \mathrm{depth}\, A$. So A is Cohen-Macaulay.

On the other hand, if A is Cohen-Macaulay, then depth $A = \dim A$. By standard techniques of homological algebra, we can show that it suffices to assume that $E^\bullet$ is a minimal injective resolution of B. Then $E^j = \coprod_{\mathrm{ht}\,\mathfrak{p}=j} E(B/\mathfrak{p})$ [Bass, 1963]. Hence $\mathrm{Ass}\, E^j = \{\mathfrak{p} : \mathrm{ht}\,\mathfrak{p} = j\}$. Since $\mathrm{Ass}\, \mathrm{Hom}_B(A, E^j) = \mathrm{Supp}_B A \cap \mathrm{Ass}\, E^j$ [Bourbaki, 1961], the set of associated prime ideals of $\mathrm{Hom}_B(A, E^j)$ is $\{\mathfrak{p} \in \mathrm{Supp}_B A : \mathrm{ht}\,\mathfrak{p} = j\}$. If A is Cohen-Macaulay, then any prime ideal in $\mathrm{Supp}_B A$ must have height at least d. Therefore, $\mathrm{Hom}_B(A, E^j) = 0$ for all $j < d$. To finish the proof, we go by induction on dim A. If dim $A = 0$, then the complex K_A is just $\mathrm{Hom}_B(A, E^n)$ and $\Omega(A) = \mathrm{Hom}_B(A, E^n)$ is injective.

Note at this point that the ring $\mathrm{Hom}_A(\Omega(A), \Omega(A))$ is just A. For $\mathrm{Hom}_A(\Omega(A), \Omega(A)) = \mathrm{Hom}_B(\Omega(A), E^n)$ by adjointness. By [Cartan-Eilenberg, 1958, Prop. VI, 5.3] this is isomorphic to $A \otimes_B \mathrm{Hom}_B(E^n, E^n)$. Now E^n is the injective envelope of the residue class field of B. Hence $\mathrm{Hom}_B(E^n, E^n)$ is just $\hat{B}$, the m-adic completion of B. But A is a B-module of finite length. Thus $A \otimes_B \hat{B}$ is just A. Since $\Omega(A)$ is injective and its endomorphism ring is A, we must have that $\Omega(A)$ is the A-injective envelope of the residue class field. Further, if A is Gorenstein, then $A = \Omega(A)$.

Suppose $d > 0$ and that we have shown the theorem to be true for all A' with $\dim A' < d$. Suppose $\dim A = d$. Then there is a regular element x in the radical of A and $\dim A/x A = -1 + \dim A$. Furthermore $A/x A$ is Cohen-Macaulay (resp. Gorenstein) if and only if A is. The exact sequence $0 \to A \xrightarrow{x} A \to A/x A \to 0$ gives rise to the long exact sequence

$$0 = \operatorname{Ext}_B^{n-d-1}(A, B) \longrightarrow \operatorname{Ext}_B^{n-d}(A/x A, B) \longrightarrow \operatorname{Ext}_B^{n-d}(A, B) \xrightarrow{x} \operatorname{Ext}_B^{n-d}(A,B)$$
$$\longrightarrow \operatorname{Ext}_B^{n-(d-1)}(A/x A, B) \longrightarrow \operatorname{Ext}_B^{n-(d-1)}(A, B) \xrightarrow{x} \operatorname{Ext}_B^{n-(d-1)}(A, B) \longrightarrow \cdots.$$

By induction $\operatorname{Ext}_B^{n-d+j}(A/x A, B) = 0$ for $j \neq 1$. Since x is in the radical of A, the surjection induced by multiplication by x on $\operatorname{Ext}_B^{n-d+j}(A, B)$ is zero for $j \neq 0$, so these groups are zero, and the sequence $0 \longrightarrow \operatorname{Ext}_B^{n-d}(A, B) \xrightarrow{x} \operatorname{Ext}_B^{n-d}(A, B) \longrightarrow \operatorname{Ext}_B^{n-(d-1)}(A/x A, B) \longrightarrow 0$ is exact. Hence x is regular on $\Omega(A)$ and $\Omega(A)/x \Omega(A) \cong \Omega(A/x A)$.

Thus the complex K_A is an injective resolution of $\Omega(A)$. And a regular A-sequence is a regular $\Omega(A)$-sequence.

Thus if x is regular in A, then $\operatorname{Hom}_A(\Omega(A), \Omega(A))/x \operatorname{Hom}_A(\Omega(A), \Omega(A))$ $\cong \operatorname{Hom}_{A/xA}(\Omega(A/x A), \Omega(A/x A))$. By induction, this module is $A/x A$. Hence $\operatorname{Hom}_A(\Omega(A), \Omega(A))$ is a free A-module of rank 1.

Also, if $A/x A$ is Gorenstein, then $\Omega(A)/x \Omega(A) = A/x A$. Hence $\Omega(A) = A$ since $\Omega(A)$ must then be free and rank one. ☐

Corollary 12.2. *Suppose A is a Cohen-Macaulay factor of a Gorenstein ring and $\Omega(A)$ is a module of dualizing differentials:*

(a) *Any regular A-sequence is a regular $\Omega(A)$-sequence.*

(b) *The set* $\{p \in \operatorname{Spec} A : A_p$ *is Gorenstein*$\} = \{p \in \operatorname{Spec} A : \Omega(A)_p$ *is projective*$\}$ *and therefore is an open set in* $\operatorname{Spec} A$. ☐

Now suppose A is a normal local ring which is a homomorphic image of the Gorenstein ring B. Then any regular A-sequence of length two is an $\Omega(A)$-sequence. Hence $\Omega(A)$ is a reflexive A-module.

Theorem 12.3 (Murthy). *Suppose the factorial local ring A is a* Cohen-Macaulay *factor ring of a Gorenstein Ring. Then A is Gorenstein.*

Proof. The module $\Omega(A)$ is a reflexive A-module. Hence it is a divisorial A-lattice in $K \otimes_A \Omega(A)$. But $K \otimes_A \Omega(A) \cong K$ since K is Gorenstein. So $\Omega(A)$ is A-isomorphic to a reflexive ideal. But any reflexive ideal is principal. Therefore $\Omega(A) \cong A$, so A is Gorenstein. ☐

Among the many applications of this theorem is the one which shows that the homogeneous coordinate ring of a Grassmannian is Gorenstein. First Samuel [1964 d] shows that these rings are factorial. Then Hochster

[1972] shows that they are Cohen-Macaulay and hence Gorenstein by Murthy's Theorem.

Many interesting properties of Cohen-Macaulay rings which are homomorphic images of Gorenstein rings are found in [Fossum, Griffith and Reiten, 1973].

Chapter III. Dedekind Domains

One of the most useful and interesting classes of Krull domains is the class of Dedekind domains. The theory is very familiar. In this chapter of three sections, special properties of Dedekind domains are discussed which are not found in the usual treatments. The first section, in addition to an introductory paragraph or two, contains a generalization of the theorem that a Krull domain with a finite number of prime ideals is a principal ideal domain. The second includes Claborn's theorem which states that any abelian group is the ideal class group of some Dedekind domain. The third section expands on this result by considering presentations of abelian groups with the intention of showing that the presentation is precisely of the form $\mathrm{Div}\,(A)/\mathrm{Prin}\,(A)$ for a Dedekind domain A.

This third section is essentially the contents of an unpublished manuscript of Claborn's. I am grateful to Mrs. Mary Lee Claborn for the opportunity to see this manuscript and for her permission to use it here.

§ 13. Dedekind Domains and a Generalized Approximation Theorem

After a few well known statements concerning Dedekind domains are given, a generalized approximation theorem for Dedekind domains is proved. The first few results are standard material whose proofs are found in the standard literature.

Theorem 13.1. *Let A be an integral domain with field of quotients K. The following statements about A are equivalent:*
  (a) *A is a Dedekind domain.*
  (b) *A is completely integrally closed and each maximal ideal is divisorial.*
  (c) *A is a Krull domain in which each nonzero prime ideal is maximal.*
  (d) *A is a Krull domain and K/A is an injective A-module.*
  (e) *Each fractionary ideal of A is divisorial.*
  (f) *Each fractionary ideal of A is invertible.*
  (g) *Each nonzero prime ideal of A is invertible.*
  (h) *A is noetherian, normal and each nonzero prime ideal is maximal.*

(i) *A is noetherian and $A_{\mathfrak{m}}$ is a principal ideal domain for each maximal ideal $\mathfrak{m}$ of A.*

(j) *A is noetherian and $A_{\mathfrak{m}}$ is a principal ideal domain for each prime ideal $\mathfrak{m}$ of A.*

(k) *A is noetherian and for each maximal ideal $\mathfrak{m}$ of A there is no ideal properly contained between $\mathfrak{m}$ and $\mathfrak{m}^2$.*

(l) *A is noetherian and the set of ideals primary for a given maximal ideal is totally ordered by inclusion.*

(m) *Each ideal of A is a product of prime ideals.*

(n) *A is a hereditary ring.*

(o) *A is noetherian and each ideal generated by two elements is projective.*

(p) *Each divisible A-module is injective.*   □

Corollary 13.2. *If A is a Dedekind domain and $A \subseteq A' \subseteq K$, then A' is a Dedekind domain.*

Proof. Since A' is torsion free as an A-module, it is flat as an A-module. Hence A' is a subintersection of A by Corollary 6.6. Therefore A' satisfies condition (c).   □

Corollary 13.3. *Each ideal in a Dedekind domain is generated by two elements.*   □

Corollary 13.4. *If A is a Dedekind domain such that $X^{(1)}(A)$ is finite, then A is a principal ideal domain.*

Proof. The approximation theorem for Krull domains gives this result immediately.   □

Theorem 13.5 (Krull-Akizuki Theorem). *Let A be a noetherian integral domain. Suppose B is a subring of a finite extension L of the field of quotients of A. If $\dim A = 1$ and B is integral over A, then B is noetherian, $\dim B = 1$ and for any nonzero a in A, the A-module B/aB is of finite length.*   □

Corollary 13.6. *If A is a noetherian domain with $\dim A = 1$, then the integral closure of A in a finite extension of its field of quotients is a Dedekind domain.*   □

The Krull-Akizuki Theorem is well known and its proofs are found in the standard references.

Claborn [1967] has found results which are analogous to Corollary 13.4. They say, in fact, that if the Dedekind domain A has enough more elements than prime ideals, then A is a principal ideal domain. The next result is a proper generalization of the corollary, provided A contains a large enough subfield.

If Z is a set, then card(Z) denotes its cardinality. We let $Z(A)$ or just Z denote the set of maximal ideals of A.

Proposition 13.7. *Let A be a Dedekind domain which contains a subfield k such that* card$(k)=$card(A). *If* card$(Z(A))<$card(A), *then A is a principal ideal domain.*

Proof. Suppose card$(Z)<$card(k). Let $\mathfrak{p}$ be a nonzero prime ideal of A. By the approximation theorem for Krull domains (5.10), choose elements f,g in $\mathfrak{p}$ with $v_\mathfrak{p}(f)=1$, $v_\mathfrak{p}(g)=2$ and such that $\mathfrak{p}=fA+gA$. If $\lambda \in k$, then $v_\mathfrak{p}(f+\lambda g)=1$. If $\mathfrak{p}$ is not principal, then for each x in $\mathfrak{p}$, and in particular for each $x=f+\lambda g$ with $\lambda \in k$, there is a $\mathfrak{q} \in Z-\{(0),\mathfrak{p}\}$ which contains x. For some such $\mathfrak{q}$ there are λ and $\lambda' \neq \lambda$, such that both $f+\lambda g$ and $f+\lambda'g$ are in $\mathfrak{q}$ since card$(k)>$card(Z) (if not then $\lambda \mapsto \mathfrak{q}_\lambda$ $\ni f+\lambda g$ is an injection of k into Z). Hence for this $\mathfrak{q}$, both g and f are in $\mathfrak{q}$ which contradicts the fact that $\mathfrak{p} \neq \mathfrak{q}$. Thus $\mathfrak{p}$ is principal. □

Now let Y be a subset of $Z(A)-\{(0)\}$. Denote by A_Y the ring $\bigcap\limits_{\mathfrak{p}\in Y} A_\mathfrak{p}$. Then A_Y is a Dedekind domain with $Z(A_Y)\cong Y$. If $\mathfrak{p}\in Y$, say $\mathfrak{p}$ is *principal with respect to Y* if the prime ideal $\mathfrak{p}A_Y$ is a principal ideal generated by an element in A. This is equivalent to saying that there is an element a in A such that $v_\mathfrak{p}(a)=1$ and $v_\mathfrak{q}(a)=0$ for all $\mathfrak{q}$ in $Y-\{\mathfrak{p}\}$.

Theorem 13.8. *Let A be a Dedekind domain. If some prime ideal of Y is not principal with respect to Y, then* card$(A)\leqslant$card$(Y)^{\aleph_0}$.

Proof. If Y is finite then the assumption is vacuous. If there is a $\mathfrak{p}$ in Z with $A/\mathfrak{p}$ finite, then card$(A)\leqslant(\text{card}(A/\mathfrak{p}))^{\aleph_0}\leqslant(\text{card}(Y))^{\aleph_0}$. Hence it may be assumed that Y is infinite and that $A/\mathfrak{p}$ is infinite for all $\mathfrak{p}$ in Z.

Lemma 13.9. *If A is a Dedekind domain in which all but a finite number of prime ideals are principal, then A is a principal ideal domain.*

Proof. Let $\mathfrak{p}_1,\ldots,\mathfrak{p}_r$ be those possibly non-principal prime ideals. Choose, for $\mathfrak{p}_1$, an element in A, say f, such that $v_{\mathfrak{p}_1}(f)=1$ and $v_{\mathfrak{p}_i}(f)=0$ for $i=2,\ldots,r$. Let $\mathfrak{q}$ be the principal ideal $g_\mathfrak{q}A$ for the other nonzero prime ideals $\mathfrak{q}$ of A. Now let $f'=f\prod\limits_\mathfrak{q} g_\mathfrak{q}^{-v_\mathfrak{q}(f)}$ where the product is taken over the prime ideals $\mathfrak{q}$ distinct from the $\mathfrak{p}_j$. Then $\mathfrak{p}_1=f'A$. □

Continuation of the proof of Theorem 13.8. Using this lemma if necessary, a set Y' can be constructed such that card$(Y)=$card(Y') and such that Y' has an infinite number of prime ideals which are not principal with respect to it. In fact, if Y contains only a finite number of such prime ideals, say $\mathfrak{p}_1, \ldots, \mathfrak{p}_r$, then a new set Y_1 is constructed as follows: For each $\mathfrak{q}$ in $Y - \{\mathfrak{p}_1, \ldots, \mathfrak{p}_r\}$ let $a_\mathfrak{q}$ be an element in $\mathfrak{q}$ which generates $\mathfrak{q} A_Y$. Let $W = \{\mathfrak{p} \in Z - Y : a_\mathfrak{q} \in \mathfrak{p} \text{ for some } \mathfrak{q}\}$. Then W is not empty by Lemma 13.9. Let $Y_1 = Y \cup W$. It is clear that each $\mathfrak{p}_j$ is not principal with respect to Y_1. But some element of W is also not principal with respect to Y_1. For suppose all such $\mathfrak{p}$ were principal with respect to Y_1. Let $a_\mathfrak{p}$ be an element in $\mathfrak{p}$ such that $\mathfrak{p} A_{Y_1} = a_\mathfrak{p} A_{Y_1}$ for each $\mathfrak{p}$ in W. By the approximation theorem, choose an element f in $\mathfrak{p}_1$ such that $v_{\mathfrak{p}_1}(f) = 1$ and $v_{\mathfrak{p}_j}(f) = 0$ for $j \neq 1$ and $v_\mathfrak{q}(f) = 0$ for $\mathfrak{q} \in Y - \{\mathfrak{p}_1, \ldots, \mathfrak{p}_r\}$. This f will have negative values for but a finite number of the $\mathfrak{p}$ in W. Let $f' = f \prod_{\mathfrak{p} \in W} a_\mathfrak{p}^{-v_\mathfrak{p}(f)}$. Then $\mathfrak{p}_1 A_{Y_1} = f' A_{Y_1}$ and $\mathfrak{p}_1$ is principal with respect to Y_1 and hence also with respect to Y, a contradiction.

Now define, inductively, $Y_n = (Y_{n-1})_1$ for $n > 1$ and let $Y' = \bigcup_{n \geq 1} Y_n$.

Then Y' has a infinite number of prime ideals which are not principal with respect to it. Thus it can be assumed that this holds for Y as well. (Certainly $\text{card}(Y) = \text{card}(Y')$ since Y' is obtained by adding but a countable number of prime ideals to Y.)

Let $\mathfrak{p}_1, \mathfrak{p}_2, \ldots$ be an infinite sequence of prime ideals which are not principal with respect to Y. Denote their respective valuations by v_i. Pick f_1 and g_1 such that $v_1(f_1) = 1$ and $v_1(g_1) = 2$ and $\mathfrak{p}_1 = f_1 A + g_1 A$. Delete from the list those prime ideals $\mathfrak{p}_j$ with $j > 1$ and $v_j(f_1) > 0$ and renumber in the original order beginning with $\mathfrak{p}_2$, etc. Choose f_2, g_2 generating $\mathfrak{p}_2$ with $v_2(f_2) = 1$ and $v_2(g_2) = 2$ and such that $g_1 f_2 \neq g_2 f_1$. Delete again those $\mathfrak{p}_j$ $(j > 2)$ which contain f_2 and renumber. Now inductively choose f_i, g_i elements in A which generate $\mathfrak{p}_j$ such that $v_j(f_j) = 1$, $v_j(g_j) = 2$ and such that: If $k < j$, then f_j/g_j is not congruent to f_k/g_k for any prime ideal $\mathfrak{q}$ for which $v_\mathfrak{q}(f_k/g_k) \geq 0$ and for which $v_\mathfrak{q}(f_m/g_m - f_n/g_n) > 0$ for m and n less than j. This is possible since $A/\mathfrak{q}$ is assumed to be infinite for all $\mathfrak{q}$ and since v_j is not positive at $f_m/g_m - f_n/g_n$ for m and n in the prescribed range. Now, as before, delete the prime ideals $\mathfrak{p}_k$ in Y which contain f_j, $1 \leq j \leq i < k$ or for which $v_k(f_m/g_m - f_n/g_n) > 0$, and renumber.

Now for each element a in A consider the elements $f_i + a g_i$. For each i it happens that $v_i(f_i + a g_i) = 1$. Since $\mathfrak{p}_i$ is not principal with respect to Y, there is a prime ideal $\mathfrak{q}_i \neq \mathfrak{p}_i$ in Y such that $f_i + a g_i$ is in $\mathfrak{q}_i$. Choose, for each i, such a $\mathfrak{q}_i$ and define $f_a : \mathbb{N} \to Y$ by $f_a(i) = \mathfrak{q}_i$. Then the image of f_a is infinite. For if not, then there is a finite number of prime ideals in Y, say $\mathfrak{r}_1, \ldots, \mathfrak{r}_m$, such that $f_i + a g_i$ is contained in one of them for each i. Hence an infinite number of $f_i + a g_i$ is contained in one of these prime ideals, say $\mathfrak{r}_1$. If $\mathfrak{r}_1$ is a $\mathfrak{p}_n$ for some n, choose $s > t > u > n$

with $f_s + ag_s$, $f_t + ag_t$, and $f_u + ag_u$ in r_1. Otherwise choose such $s > t > u$. Now $g_j \notin r_1$ for $j = s, t, u$ for otherwise $r_1 = p_j$, a contradiction. Hence g_j is invertible modulo r_1 for $j = s, t, u$ and $f_s/g_s \equiv f_t/g_t \pmod{r_1}$. Now $v_{r_1}(f_t/g_t - f_u/g_u) > 0$ which contradicts the construction of the f_j and g_j. This shows that the image of f_a is infinite.

Thus $a \mapsto f_a$ defines a function $A \to Y^{\mathbf{N}}$. To establish the theorem, it is enough to show that this function is an injection. Suppose that $f_a = f_b$ for a and b in A. Then $f_i + ag_i$ and $f_i + bg_i$ are in the same ideal q_i for all i. Hence $(a - b)g_i$ is in q_i for all i. But g_i is not in q_i since $q_i \neq p_i$. Hence $a - b$ is in q_i for all i. Since the set q_i is infinite, $a = b$ (since $v_{q_i}(a - b) = 0$ for all but a finite number of q_i). $\square$

There now follow two corollaries, one an analogue of the theorem that a Dedekind domain with a finite number of prime ideals is a principal ideal domain and the other a generalization of the approximation theorem.

Corollary 13.10. *If A is a Dedekind domain with $(\mathrm{card}(Z))^{\aleph_0} < \mathrm{card}(A)$, then A is a principal ideal domain.* $\square$

Corollary 13.11 (Generalized approximation theorem). *Suppose A is a Dedekind domain and that Y is a subset of $Z(A)$. If $\mathrm{card}(Y)^{\aleph_0} < \mathrm{card}(A)$ and if $f \in \mathbf{Z}^{(Y)}$, there is an element a in A such that $v_p(a) = f(p)$ for all p in Y.* $\square$

§ 14. Every Abelian Group is an Ideal Class Group

The main purpose of this section is to establish Claborn's Theorem [1966a] that every abelian group is isomorphic to the divisor class group of some Dedekind domain.

There are three basic parts to the demonstration. One part is to show that a Krull domain A admits an extension B which is a Dedekind domain and such that their class groups are isomorphic. Another part is to exhibit a Krull domain B which is an extension of A such that the class group of B is the class group of A modulo some given subgroup. The remaining part is to construct a Krull domain whose class group is isomorphic to the free abelian group based on some preassigned set.

This first result is technical, but important for the remainder of this section.

Lemma 14.1. *Let A be an integral domain and a and b relatively prime elements in A (i.e., $Aa \cap Ab = Aab$). Then $bX - a$ is a prime element in $A[X]$.*

Proof. Consider the homomorphism $A[X] \to K$, the quotient field of A, given by $g \mapsto g(a/b)$. If it can be demonstrated that the kernel is the principal ideal generated by $bX - a$ then the conclusion will follow. Suppose $g(a/b) = 0$. We will show that $g \in (bX - a)$ by induction on the degree of g. Let $g = a_0 X^n + a_1 X^{n-1} + \cdots + a_n$. Then $a_0 a^n + a_1 a^{n-1} b + \cdots + a_n b^n = 0$. Hence $a_0 a^n \in a^n A \cap bA = a^n bA$. Thus $a_0 \in bA$, say $a_0 = a_0' b$. Then $g = a_0'(bX - a) X^{n-1} + (a_0' a + a_1) X^{n-1} + \cdots + a_n$. Now $g - \{a_0'(bX - a) X^{n-1}\}$ is a polynomial of smaller degree than g which is zero on a/b. Hence it is divisible by $bX - a$. Thus g is divisible by $bX - a$. $\Box$

Now the first important result can be established.

Theorem 14.2. *Let A be a Krull domain. Then there is a flat extension $A \to B$ satisfying the properties:*

(a) *B is a Dedekind domain.*

(b) *If v is an essential rank one discrete valuation of A, then there is one and only one extension of v to an essential valuation of B with index of ramification one (B is an inert extension in the terminology of Samuel [1968]).*

(c) *$\mathrm{Cl}(A) \to \mathrm{Cl}(B)$ is a bijection.*

Proof. Let $R = A[X_1, \ldots, X_n, \ldots]$ be a transcendental extension of A of countable degree. It is clear that R is a Krull domain, that $A \to R$ is flat, and that the extension satisfies properties (b) and (c) of the theorem.

Suppose $\mathfrak{Q}$ is a prime ideal in R with $\mathrm{ht}\,\mathfrak{Q} > 1$. Let q be a nonzero element in $\mathfrak{Q}$ and suppose that $\mathfrak{P}_1, \ldots, \mathfrak{P}_k$ are the prime divisorial ideals of R which contain q. Then there is a q' in $\mathfrak{Q} - (\mathfrak{P}_1 \cup \cdots \cup \mathfrak{P}_k)$. The elements q and q' are relatively prime. Let $X_\mathfrak{Q}$ be a variable occurring in neither q nor q' and form $f_\mathfrak{Q} = q X_\mathfrak{Q} - q'$. Then $f_\mathfrak{Q}$ is a prime element in R by the previous lemma.

Let S be the multiplicatively closed subset of R generated by the prime elements $f_\mathfrak{Q}$, $\mathrm{ht}\,\mathfrak{Q} > 1$. Let B denote the ring $S^{-1} R$. Now $A \to B$ is flat and B is a Krull domain in which every prime ideal distinct from zero is maximal. Hence B is a Dedekind domain. Thus condition (a) is satisfied. By Nagata's Theorem $\mathrm{Cl}(R) \to \mathrm{Cl}(B)$ is an isomorphism, so condition (c) is fulfilled.

The extension $A \to R$ is clearly inert. The only discrete rank one valuations which get lost in going from R to B are those whose centers on R are principal ideals generated by polynomials of the form $q X - q'$. These can never be extensions of valuations of A. Hence the total extension $A \to B$ is inert. $\Box$

One question which can be raised at this point is: What is the "minimal" such extension? Clearly, if A is itself Dedekind, then $B = A$ works. This question has to my knowledge not been investigated.

Let $\mathscr{Q}$ be a collection of divisorial prime ideals of the Krull domain A. Let $C(\mathscr{Q})$ denote the subgroup of $Cl(A)$ generated by the classes in $Cl(A)$ of the elements in $\mathscr{Q}$. Let $B = \bigcap_{\mathfrak{p} \in \mathscr{Q}} A_{\mathfrak{p}}$. The surjection $Cl(A) \to Cl(B)$ has kernel $C(\mathscr{Q})$ by Nagata's Theorem.

If H is some subgroup of $Cl(A)$, it is not the case that $H = C(\mathscr{Q})$ for some set $\mathscr{Q}$. In another section of this chapter a Dedekind domain will be constructed for which the class group is finite cyclic and the class of every prime ideal is in the generating class. Such a Dedekind domain satisfies the condition $C(\mathscr{Q}) = Cl(A)$ for any nonempty subset $\mathscr{Q}$ of the set of nonzero prime ideals. Since the object is to form a Krull domain B with $Cl(B) = Cl(A)/H$, it is seen that a subintersection does not answer immediately to the question. Thus the next result is crucial.

Theorem 14.3. *Let A be a Krull domain. Then every divisor class of $A[X]$ contains a prime divisorial ideal.*

Proof. The map $A \to A[X]$ induces a bijection $Cl(A) \to Cl(A[X])$. So there is an ideal of the form $\mathfrak{a}A[X]$, with $\mathfrak{a}$ divisorial in A, in every class of $Cl(A[X])$. We would like to show that there is a prime ideal in $cl(\mathfrak{a}A[X])$. By Proposition 5.11, there are elements a_1 and a_0 both different from zero and in $A:\mathfrak{a}$ such that $\mathfrak{a} = A a_0^{-1} \cap A a_1^{-1}$. The ideal in $K[X]$ generated by the linear $a_1 X - a_0$ is a prime ideal. Denote its intersection with $A[X]$ by $\mathfrak{P}$. The claim is that $\mathfrak{P} = (a_1 X - a_0)\mathfrak{a}A[X]$. If this is indeed the case, then $cl(\mathfrak{P}) = cl(\mathfrak{a}A[X])$, so the prime divisorial ideal is in $cl(\mathfrak{a}A[X])$. So the proof will be complete when this equality has been demonstrated.

We now show that $\mathfrak{P} = (a_1 X - a_0)\mathfrak{a}A[X]$. Since $a_i \mathfrak{a} \subseteq A$ for each i, the ideal $(a_1 X - a_0)\mathfrak{a}A[X]$ is in $A[X]$ and in $(a_1 X - a_0)K[X]$, so $\mathfrak{P} \supseteq (a_1 X - a_0)\mathfrak{a}A[X]$. Suppose that the polynomial $g \in \mathfrak{P}$. Then $g = (a_1 X - a_0)h$ for some h in $K[X]$. To show that $\mathfrak{P} \subseteq (a_1 X - a_0)\mathfrak{a}A[X]$, it is sufficient to show that $h \in \mathfrak{a}A[X]$. We do so by induction on the degree of h.

If $\deg h = 0$, then h is a nonzero constant in $K[X]$. Then $(a_1 X - a_0)h \in A[X]$ if and only if $a_1 h \in A$ and $a_0 h \in A$. But then $(a_1 X - a_0)h \in A[X]$ if and only if $h \in \mathfrak{a}$. Now $h \in \mathfrak{a}A[X]$ if and only if the coefficients of h are in $\mathfrak{a}$. Suppose that $h_1 \in \mathfrak{a}A[X]$ for all polynomials h_1 with $0 \leqslant \deg h_1 < \deg h$ such that $(a_1 X - a_0)h_1 \in A[X]$. Suppose further that $(a_1 X - a_0)h \in A[X]$. To show that $h \in \mathfrak{a}A[X]$, it is sufficient to show that the leading coefficient of h is in $\mathfrak{a}$. Let $h = b_d X^d + b_{d-1} X^{d-1} + \cdots + b_0$ with $b_d \neq 0$. Then $b_d \in \mathfrak{a}$ if and only if $b_d a_0 \in A$ and

$b_d a_1 \in A$. Now $(a_1 X - a_0)h = (a_1 b_d) X^{d+1} + \sum\limits_{n=d}^{1} (a_1 b_{n-1} - a_0 b_n) X^n - a_0 b_0$.
Since $(a_1 X - a_0)h \in A[X]$, the coefficients $a_1 b_d$, $a_0 b_0$ and $a_1 b_{n-1} - a_0 b_n$, $d \geqslant n \geqslant 1$, are all in A. Thus $b_d \in \mathfrak{a}$ if and only if $a_0 b_d \in A$. And $a_0 b_d \in A$ if and only if $a_0 b_d$ is integral over A. We now proceed to establish that $a_0 b_d$ is integral over A.

If $d = 1$, then the coefficients are $a_1 b_1$, $a_1 b_0 - a_0 b_1$ and $a_0 b_0$. Now $(a_0 b_1)^2 - (a_1 b_0 - a_0 b_1)(a_0 b_1) - (a_1 b_1)(a_0 b_0) = 0$ so $(a_0 b_1)$ is integral over A. For general d an equation of integrality is $(a_0 b_d)^{d+1}$
$- \sum\limits_{n=d}^{1} [(b_d a_1)^{d-n}(a_0 b_n - a_1 b_{n-1})(a_0 b_d)^n] - (a_1 b_d)^d a_0 b_0 = 0$. This is seen to be the case by noting that the sum can be written as

$$\sum_{n=d}^{1} b_d^d (a_0^{n+1} a_1^{d-n} b_n - a_0^n a_0^{d-n+1} b_{n-1})$$

and that in this sum all the terms cancel except the term $(a_0 b_d)^{d+1}$ coming from $n = d$ and the term $-(a_1 b_d)^d a_0 b_0$ coming from $n = 1$. Hence $a_0 b_d$ is integral over A and so is in A. Thus $b_d \in \mathfrak{a}$ and the inductive bridge is completed.   ☐

Corollary 14.4. *If H is a subgroup of* $\mathrm{Cl}(A)$, *there is an extension* $A \to B$ *such that* $0 \to H \to \mathrm{Cl}(A) \to \mathrm{Cl}(B) \to 0$ *is an exact sequence.*

Proof. Let $R = A[X]$. Let H be the image in $\mathrm{Cl}(R)$ of H. For each h in H, let $\mathfrak{P}_h$ be a prime divisorial ideal in R such that $\mathrm{cl}(\mathfrak{P}_h) = h$. Let $\mathcal{Q} = \{\mathfrak{P}_h : h \in H\}$. Then $C(\mathcal{Q}) = H$. Now take for B the subintersection $B = \bigcap R_{\mathfrak{P}}$, $\mathfrak{P} \in \mathcal{Q}$. The result follows from Nagata's Theorem.   ☐

If G is a given abelian group, then G can be presented by a free abelian group modulo a subgroup: $G = \mathbb{Z}^{(S)}/H$. The goal now is to find a Krull domain whose class group is $\mathbb{Z}^{(S)}$. Then by the corollary above and Theorem 14.2, there is a Dedekind domain with class group G.

A construction of a Krull domain A with divisor class group isomorphic to $\mathbb{Z}^{(S)}$, the free abelian group based on a set S, is deduced from the affine ring of the hypersurface $xv - yu = 0$. By taking enough copies (S of them) of this quadratic over a field, a Krull domain with divisor class group $\mathbb{Z}^{(S)}$ is obtained.

In what follows, lower case x, y, u, v will denote elements in a ring subject to the relation $xv - yu = 0$. Upper case X, Y, U, V will denote indeterminates.

Proposition 14.5. *Let A be a Krull domain.*
(a) *Then $A[x, y, u, v]$ is a Krull domain.*
(b) $A[x, y, u, v] = A[x, y, u/x] \cap A[x, y, u, v]_{(x,y)}$ *where $x, y, u/x$ are al-*

gebraically independent over the field of quotients of A and (x, y) is a prime divisorial ideal.

(c) *This representation is irredundant.*

The ring $A[x,y,u,v]$ is an integral domain by Lemma 14.1. The proof follows the next lemma.

Lemma 14.6. *Let C be an integral domain with field of quotients K.*

(a) *Then the ideal in $C[x,y,u,v]$ generated by x and y is a prime ideal.*

(b) *The local ring $C[x,y,u,v]_{(x,y)}$ is a principal valuation ring.*

(c) *Finally, $C[x,y,u,v] = C[x,y,u/x] \cap C[x,y,u,v]_{(x,y)}$.*

Proof. (a) The residue ring $C[x,y,u,v]/(x,y) \cong C[X,Y,U,V]/(X,Y,XV-YU) \cong C[U,V]$ which is an integral domain. Therefore (x,y) is a prime ideal.

(b) Let $\mathfrak{X} = (X,Y)$ in $D = C[X,Y,U,V]$. Then $C[x,y,u,v]_{(x,y)} = D_{\mathfrak{X}}/(XV-YU)D_{\mathfrak{X}}$. Now $D_{\mathfrak{X}} = K(U,V)[X,Y]_{(X,Y)}$ is a regular local ring. In this local ring $(XV-YU)D_{\mathfrak{X}} = (X/U - Y/V)D_{\mathfrak{X}}$. But $X/U - Y/V$ is an element not in the square of the maximal ideal. Hence the residue ring is again a regular local ring. It is of dimension one so a principal valuation ring.

(c) Since the transcendence degree of $K(x,y,u,v)$ over K is 3 and since $K(x,y,u,v) = K(x,y,u/x)$, it follows that $x,y,u/x$ are algebraically independent over K. Let f be an element in $C[x,y,u/x] \cap C[x,y,u,v]_{(x,y)}$. We want to show that $f \in C[x,y,u,v]$. Write $f = c_0 + \cdots + c_n(u/x)^n$ where $n \geqslant 0$ and $c_j \in C[x,y,u,v]$ for each j. Suppose, for inductive purposes, that any g in $C[x,y,u/x] \cap C[x,y,u,v]_{(x,y)}$ whose degree in u/x is less than n is in $C[x,y,u,v]$. Now there are elements b and s in $C[x,y,u,v]$, with $s \notin (x,y)$, such that $f = b/s$. Then $x^n b = s c_0 x^n + s c_1 x^{n-1} u + \cdots + s c_n u^n$, so $s c_n u^n \in (x,y)$. Since $s u^n \notin (x,y)$, the element $c_n \in (x,y)$. Write $c_n = a x + c y$ with a and c in $C[x,y,u,v]$. Then $c_n u u^{n-1} = (a x u + c y u) u^{n-1} = (a u + b v) x u^{n-1}$. Hence $f = c_0 + c_1(u/x) + \cdots + (c_{n-1} + a u + c v)(u/x)^{n-1}$. By induction $f \in C[x,y,u,v]$. □

Proof of Proposition 14.5. If A is a Krull domain, then property (c) of the lemma shows that $A[x,y,u,v]$ is the intersection of two Krull domains, so it is a Krull domain. In fact, the principal valuation rings of $A[x,y,u/x]$ and the valuation ring $A[x,y,u,v]_{(x,y)}$ are the essential valuations of $A[x,y,u,v]$, i.e., the representation of $A[x,y,u,v]$ as the intersection of these two Krull domains is irredundant. To see this, it is enough to show that to each prime divisorial ideal of $A[x,y,u,v]$ there is associated an element in $K(x,y,u,v)$ which is not in the associated valuation ring, but lies in all other valuation rings. Now $u/x \in A[x,y,u/x]$ but it is not in $A[x,y,u,v]_{(x,y)}$. Suppose $\mathfrak{P}$ is a divisorial prime ideal in

$A[x,y,u/x]$. If $\mathfrak{P}=(x)$, then y/x has valuation -1 at $\mathfrak{P}$, at least zero for all other prime ideals of $A[x,y,u/x]$ and has value zero in $A[x,y,u,v]_{(x,y)}$. If $\mathfrak{P}$ is distinct from $xA[x,y,u/x]$, by the approximation theorem, pick f in $K(x,y,u,v)$ with value -1 at $\mathfrak{P}$ and positive for all other prime divisorial ideals of $A[x,y,u/x]$. If $v_{(x,y)}(f)<0$, then $v_{(x,y)}(x^r f)\geqslant 0$ for some large enough r. But x^r is a unit at $\mathfrak{P}$ and has non-negative values at the other prime ideals. This shows that $\mathfrak{P}$ is essential. Now all the properties of Proposition 14.5 are established.   □

Corollary 14.7. *Let A be a Krull domain, S a set, $\{X_s, Y_s, U_s, V_s\}_{s\in S}$ a collection of indeterminates. Let $A^{(S)} = A[\{X_s, Y_s, U_s, V_s\}_{s\in S}]/(\{X_s V_s - Y_s U_s\}_{s\in S})$. Then*
  (a) *$A^{(S)}$ is a Krull domain.*
  (b) *$A^{(S)} = A[\{x_s, y_s, u_s/x_s\}_{s\in S}] \cap \bigcap\limits_{s\in S} A^{(S)}_{(x_s,y_s)}$.*
  (c) *The intersection is irredundant.*

Proof. An element in $A^{(S)}$ involves only a finite number of the indeterminants. Thus (a) and (b) follow immediately. The proof of (c) is almost the same as the proof in the proposition.   □

Proposition 14.8. *If K is a field and S a set, then $\mathrm{Cl}(K^{(S)})=\mathbb{Z}^{(S)}$.*

Proof. Let $B=K[\{x_s,y_s,u_s,v_s\}_{s\in S}]$ and $C=K[\{x_s,y_s,u_s/x_s\}_{s\in S}]$. Then $\mathrm{Cl}(C)=(0)$ since C is a purely transcendental extension of K. The ring C is a subintersection of B, so $\mathrm{Ker}(\mathrm{Cl}(B)\to\mathrm{Cl}(C))$ is generated by those prime ideals omitted when C is defined. Thus $\mathrm{Cl}(B)$ is generated by those prime divisorial ideals (x_s,y_s), $s\in S$.

Suppose there is a relation among their classes, say, an element f in the field of quotients of B such that $\mathrm{div}(f B)= \sum\limits_{s\in S} n_s \mathrm{div}(x_s,y_s)$ with almost all $n_s=0$. Then $f B= \bigcap\limits_{s\in S}(x_s,y_s)^{(n_s)}$. Now $f C=C$, so f is a unit in C. The only units in C are the units in K. Hence f is a unit in B and the relation is the trivial one. Thus the classes freely generate $\mathrm{Cl}(B)$.   □

Corollary 14.9. *Let A be a Krull domain with field of quotients K. Let S be a set. Then $\mathrm{Cl}(A^{(S)}) = \mathrm{Cl}(A)\oplus\mathbb{Z}^{(S)}$.*

Proof. Consider the extension $A\to K$. It induces the extension $A^{(S)}\to K^{(S)}$ which in turn induces an epimorphism $\mathrm{Cl}(B)\to\mathbb{Z}^{(S)}$. The kernel is generated by the classes of the prime divisorial ideals which meet $A-(0)$, i.e., the extensions of the prime divisorial ideals of A.   □

By suitably modifying the proof it should be clear that, given A and an abelian group G, it is possible to find an extension B of A such that the group $\mathrm{Cl}(B)$ is the split extension of G by $\mathrm{Cl}(A)$. This leads to another question: Is every element of $\text{Ext}^1_{\mathbb{Z}}(G, \text{Cl}(A))$ represented by an exact sequence of the form $0 \to \text{Cl}(A) \to \text{Cl}(B) \to G \to 0$?

Now the main result of Claborn follows.

Theorem 14.10. *Let G be an abelian group. There is a Dedekind domain D such that $\text{Cl}(D) \cong G$.*

Proof. The theorem follows from Theorems 14.2, 14.3, and 14.8. $\square$

We can take a more complicated ring for our example of a Krull domain with class group $\mathbb{Z}$. Actually, the ring $k[x, y, u, v]$ is the homogeneous coordinate ring for the Segre embedding $\mathbf{P}^1 \times \mathbf{P}^1 \to \mathbf{P}^3$. If one considers the Segre embedding $\mathbf{P}^m \times \mathbf{P}^1 \to \mathbf{P}^{2m+1}$, the resulting ring is $k[x_1, \ldots, x_m, y_1, \ldots, y_m]$ with $x_i y_j = x_j y_i$ for $1 \leqslant i < j \leqslant m$. Of course $k[x, y, u, v]$ is just the case $m=2$. More generally, one can consider the ring $A[x_1, \ldots, x_m, y_1, \ldots, y_m]$ where A is an integral domain and $x_i y_j = x_j y_i$ for the same values of i and j. According to Northcott [1963], this ring is again an integral domain. Its Krull dimension is $m + 1 + \dim A$. Just as in the case $m=2$, we can prove the following results.

Proposition 14.11.. *Let A be a Krull domain. Then*

(a) $\mathfrak{p} = (x_1, \ldots, x_m)$ *is a prime ideal and $A[x_1, \ldots, x_m, y_1, \ldots, y_m]_{\mathfrak{p}}$ is a principal valuation ring.*

(b) $A[x_1, \ldots, x_m, y_1, \ldots, y_m] = A[x_1, \ldots, x_m, y_j/x_i] \cap A[x_1, \ldots, y_m]_{(x_1, \ldots, x_m)}$ (this holds for any domain).

(c) *This is an irredundant representation of $A[x_1, \ldots, x_m, y_1, \ldots, y_m]$ as the intersection of principal valuation rings and it is a Krull domain.*

(d) $\text{Cl}(A[x_1, \ldots, x_m, y_1, \ldots, y_m]) = \text{Cl}(A) \oplus \mathbb{Z}$. $\square$

By taking large enough products of the ring $A[x_1, \ldots, y_m]$ we can get a ring $A^{(S)}$ whose class group is $\text{Cl}(A) \oplus \mathbb{Z}^{(S)}$. We remark that J. Sally [1971] has generalized some of the material concerning quadratic transforms which appears in this section.

Another remark is appropriate here. According to Corollary 10.3, the group $\text{Cl}(A) = \text{Cl}(A_{\mathfrak{m}})$ where $\mathfrak{m}$ is the maximal homogeneous ideal. Thus $k[x, y, u, v]_{\mathfrak{m}}$ is a local ring with class group $\mathbb{Z}$.

Theorem 14.10 gives an answer to Remark (3) in Rosenberg and Zelinsky [1961]. Let A be a Dedekind domain Λ a central separable A-algebra (an Azumaya algebra in the current terminology). Let $J(A)$ denote the set of left A-isomorphism classes of left $\Lambda \otimes_A \Lambda^0$-modules P with the property that $P_{\mathfrak{p}} \cong \Lambda_{\mathfrak{p}}$ for all $\mathfrak{p}$ in $Z(A)$. The question mentioned above asks whether $J(A) = 0$ for all A implies $\text{Cl}(A) = 0$. When $\text{Cl}(A)$ is torsion free, it follows from Theorem 15 [loc. cit.] that $J(\Lambda) = 0$. Hence the answer is negative, in general.

Prill [1971] has shown that any finitely generated abelian group is a divisor class group of a normal analytic local ring (which is noetherian).

Further, he shows that each group $\mathbb{Q}^c \oplus (\mathbb{Q}/\mathbb{Z})^N$ $(0 \leqslant N \leqslant \infty)$ is the divisor class group of a normal analytic local ring. Suppose H is a finite cyclic subgroup of $GL(2,\mathbb{C})$ whose generator fixes only $(0,0)$. Then the local ring at the singular point of $\mathbb{C}^2/H$ has divisor class group H. Prill describes how to form a ring whose divisor class group is the sum of a finite number of other class groups. Using the above local ring and the local ring of the cone of the Segre embedding $\mathbf{P}^1 \times \mathbf{P}^1 \to \mathbf{P}^3$, he can then get any finitely generated abelian group as the class group of a local ring. His methods are mostly geometric and are completely outside the scope of this book.

A related question is posed by Kaplansky. Let A be a noetherian local integral domain. The Grothendieck group $K^0(A)$ of finitely generated A-modules has a decomposition $K^0(A) = \mathbb{Z} \oplus \tilde{K}^0(A)$. Can $\tilde{K}^0(A)$ be any abelian group?

§ 15. Presentations of Ideal Class Groups of Dedekind Domains

In the previous section we gave Claborn's construction of a Dedekind domain having a specified abelian group as its class group. In this section, we report a result, due also to Claborn, concerning construction of Dedekind rings with very precise information about the subgroup of principal divisors. Claborn [1968] reported a countable version of the material below, and promised an uncountable version later. A manuscript containing this uncountable version has been found (thanks to his wife who was very cooperative in the search). The account which follows is essentially that found in this manuscript. The techniques involve transfinite induction, which we assume is familiar. The reader is referred to Claborn [loc. cit.] for a countable version, which is very similar, and which provides an outline of what follows. The basic idea is to find a Dedekind domain A with $Z(A) \cong X$, where X is a given set, in such a way that the subgroup of principal divisors of the free abelian group on X has very specific properties.

In order to best expose the results, it is necessary first to establish some notation. Let X be a set. The free abelian group based on X is denoted by $\mathbb{Z}^{(X)}$. The elements we think of as the functions $f: X \to \mathbb{Z}$ which have finite support. We denote by $\mathbb{Z}_+^{(X)}$ those functions which take non-negative integers as values. This set is obviously equivalent with $\mathbb{N}^{(X)}$, but we will prefer the former to the latter for technical reasons. In particular, if U is a subset of $\mathbb{Z}^{(X)}$, we will denote by U_+ the set $U \cap \mathbb{Z}_+^{(X)}$. The domain space X is embedded in $\mathbb{Z}^{(X)}$ by the coordinate maps. That is, if x and y are in X, then $x(y) = \delta_{xy}$ defines the function x. We will not distinguish the elements X from their functions.

If X' is a subset of X, then restriction defines a group homomorphism $\rho: \mathbb{Z}^{(X)} \to \mathbb{Z}^{(X')}$ which is a surjection. It also defines a surjection $\rho: \mathbb{Z}_+^{(X)} \to \mathbb{Z}_+^{(X')}$. It will not be convenient to adorn the function ρ with super- or subscripts.

Let P be a subset of $\mathbb{Z}_+^{(X)}$. We say P is *finitely dense* in $\mathbb{Z}^{(X)}$ if $\rho(P) = \mathbb{Z}_+^{(X')}$ for every finite subset $X' \subset X$. We say P is *dense* in $\mathbb{Z}^{(X)}$ if $\rho(P) = \mathbb{Z}_+^{(X')}$ for every subset X' such that $\mathrm{card}(X') < \mathrm{card}(X)$. Clearly, if P is finitely dense, every element in $\mathbb{Z}^{(X)}$ can be approximated by an element in $\mathrm{Gr}(P)$, the subgroup generated by P. Also, if P is dense, it is finitely dense.

The approximation theorem for Krull domains (Proposition 5.8) states that $(\mathrm{Prin}\, A)_+$ is finitely dense in $\mathrm{Div}(A)$ (where $\mathrm{Div}\, A = \mathbb{Z}^{(X^1(A))}$).

The generalized approximation theorem for Dedekind domains, Corollary 13.11, has the following formulation: Let $X' \subset Z(A)$ be a subset such that $(\mathrm{card}(X'))^{\aleph_0} < \mathrm{card}(A)$. Then $\rho(\mathrm{Prin}\, A)_+ = \mathbb{Z}_+^{(X')}$.

Note that if $Y \subset Z(A)$, then $\rho(\mathrm{Prin}_+ A)_+ = \mathbb{Z}_+^{(Y)}$ if and only if every prime in Y is principal with respect to Y. (See the previous section for the definition of the terminology "principal with respect to".) Also, to say that $(\mathrm{Prin}\, A)_+$ is dense in $\mathrm{Div}\, A$ is equivalent to saying that each subintersection $\bigcap_{\mathfrak{p} \in Y} A_\mathfrak{p} = A_Y$ is factorial for all $Y \subset Z(A)$ such that $\mathrm{card}(Y) < \mathrm{card}(A)$.

Now the question, which will be partially answered below, is this: Let X be a set and P a subset which is dense in $\mathbb{Z}^{(X)}$. Is there a Dedekind domain A and a bijection $\sigma: X \to Z(A)$ such that $\sigma(\mathrm{Prin}\, A) = \mathrm{Gr}(P)$ (where we denote also by σ the associated isomorphism $\mathbb{Z}^{(Z(A))} \cong \mathrm{Div}\, A \overset{\sigma}{\cong} \mathbb{Z}^{(X)}$)?

If X is countable, then finitely dense is equivalent to dense. Claborn [loc. cit.] has answered this question affirmatively in this case. However, in the uncountable case there is a difficulty encountered because, while $(\mathrm{Prin}\, A)_+$ is always finitely dense, it is not always dense. On the other hand, there are finitely dense P which can never be the subgroup of principal divisors. Examples supporting these claims appear below.

When (A, σ) satisfies the property of the question we will say that (A, σ) (or just A) *realizes the pair* (X, P). That is, if P is a (finitely) dense subset of $\mathbb{Z}^{(X)}$ and there is a bijection $\sigma: X \to Z(A)$ such that the induced isomorphism $\sigma: \mathrm{Div}\, A \cong \mathbb{Z}^{(X)}$ induces a bijection $\sigma: \mathrm{Prin}\, A \to \mathrm{Gr}(P)$, then we say that (A, σ) *realizes the pair* (X, P).

The first result is a negative one, showing that one must assume more than finite denseness in order to have the hope of realizing a pair.

Proposition 15.1. *Let X be a set with $\mathrm{card}(X) \geqslant 2^c$. There is a finitely dense P such that the pair (X, P) cannot be realized.*

Proof. Since $\operatorname{card}(X) \geqslant 2^c$, the set X can be partitioned into $\operatorname{card} X$ subsets each having a countable number of elements. We can index these blocks by X; so $X = \bigcup_{z \in X} B_z$ with $\operatorname{card}(B_x) = \aleph_0$ and such that this is a partition. Let $P = \left\{ f \in \mathbb{Z}^{(X)} : \sum_{x \in B_z} f(x) \text{ is even for all } z \in X \right\}$. Then P clearly is finitely dense in $\mathbb{Z}^{(X)}$.

Suppose (A, σ) realizes (X, P). We identify $Z(A)$ with X via σ. Choose c elements in A and let R be the ring generated by these elements. Then $\operatorname{card}(R) = c$. Each of the c nonzero elements in R lies in only a finite number of prime ideals in A. Thus there are at most c prime ideals in $Z(A)$ which contain nonzero elements of R. Hence there must be a whole block of prime ideals whose elements intersect R in (0). Let B be one such block.

Let $\mathfrak{p}$ be in B. Choose f and g in $\mathfrak{p}$ such that $\mathfrak{p} = (f, g)$, with $v_\mathfrak{p}(f) = 1$, and $v_\mathfrak{p}(g) = 2$ (by the approximation theorem). Consider the c elements $f + rg$ as r runs through R. Then $v_\mathfrak{p}(f + rg) = 1$ for each of these elements. Since A realizes the pair (X, P) and $v_\mathfrak{p}(f + rg) = 1$, there is at least one other prime ideal $\mathfrak{q}$ in B such that $v_\mathfrak{q}(f + rg) > 0$ for each r in R. But there are c such elements and only $\aleph_0$ available prime ideals to fit them into. Thus there is a prime ideal $\mathfrak{q}$ in B, $\mathfrak{q} \neq \mathfrak{p}$, which contains two distinct such elements, say $f + rg$ and $f + r'g$. But then $(r - r')g \in \mathfrak{q}$. But $r - r' \notin \mathfrak{q}$ so $g \in \mathfrak{q}$. As $f + rg \in \mathfrak{q}$, it follows that both f and g are in $\mathfrak{q}$ and so $\mathfrak{q} = \mathfrak{p}$, a contradiction to the assumption that A realizes (X, P). $\square$

One result used in the proof, and which is necessary for future use, is the relation between $\operatorname{card} A$ and $\operatorname{card} Z(A)$. We record below several set theoretical results which will be needed.

Lemma 15.2. *Let A be a Dedekind domain. Then* $\operatorname{card} A \geqslant \operatorname{card} Z(A)$.

Proof. Clearly $\operatorname{card} A = \operatorname{card} K$, where K is the field of quotients of A. For each $\mathfrak{p}$ in $Z(A)$ choose, by the approximation theorem, an element x in K such that $v_\mathfrak{p}(x) = -1$ and $v_\mathfrak{q}(x) \geqslant 0$ for all $\mathfrak{q} \neq \mathfrak{p}$. This defines a function $Z(A) \to K$ which is clearly an injection. $\square$

A corresponding lemma can be stated for a pair (X, P) where P is finitely dense.

Lemma 15.3. *Let P be finitely dense in $\mathbb{Z}^{(X)}$. Then* $\operatorname{card} P = \operatorname{card} \mathbb{Z}^{(X)}$.

Proof. Let P' be the subgroup generated by P. Then $\operatorname{card} P' = \operatorname{card} P$. If $X' \subset X$, where X' is finite, then $\rho(P') = \mathbb{Z}^{(X')}$. For each x in X there is an element f in P' such that $f(x) = -1$ and $f(x') \geqslant 0$ for all $x' \in X - \{x\}$. This defines a function $X \to P'$ which is clearly an injection. Hence $\operatorname{card} X \leqslant \operatorname{card} P$. If X is finite itself, then $P' = \mathbb{Z}^{(X)}$. Otherwise $\operatorname{card} \mathbb{Z}^{(X)} \leqslant \operatorname{card} P$. $\square$

Corollary 15.4. *If P is dense in $\mathbb{Z}^{(X)}$, then* $\operatorname{card} P = \operatorname{card} \mathbb{Z}^{(X)}$.

Proof. If P is dense, then P is finitely dense. □

Lemma 15.5. *Let A be a Dedekind domain and $\mathfrak{p}$ a maximal ideal. Then* $\operatorname{card} A \leqslant (\operatorname{card} A/\mathfrak{p})^{\aleph_0}$.

Proof. The Dedekind domain A is embedded in the $\mathfrak{p}$-adic completion $\hat{A}_\mathfrak{p} = \varprojlim_n A/\mathfrak{p}^n$. Hence $\operatorname{card} A \leqslant \operatorname{card} \hat{A}_\mathfrak{p} \leqslant (\operatorname{card} A/\mathfrak{p})^{\aleph_0}$. □

Lemma 15.6. *Let A be a Dedekind domain. Let γ denote* $\sup\{\operatorname{card} A/\mathfrak{p} : \mathfrak{p} \in Z(A)\}$. *If $\gamma > \operatorname{card} A/\mathfrak{p}$ for all $\mathfrak{p}$ in $Z(A)$, then* $\operatorname{card} Z(A) \leqslant \gamma$.

Proof. Since $\operatorname{card} A/\mathfrak{p} \leqslant \operatorname{card} A$ for all $\mathfrak{p}$ in $Z(A)$, then $\operatorname{card} A \geqslant \gamma$. Hence there is a subset $T \subseteq A$ such that $\operatorname{card} T = \gamma$. Since $\operatorname{card} T > \operatorname{card} A/\mathfrak{p}$ for each $\mathfrak{p}$ in $Z(A)$ by assumption, there are two elements t, t' in T, $t \neq t'$, whose difference lies in $\mathfrak{p}$. There are γ such differences, each lying in a finite number of prime ideals. Hence $\operatorname{card} Z(A) \leqslant \gamma$. □

The next result, purely set theoretical in nature, is used in the construction of a Dedekind domain whose principal divisors are not dense in the divisor class group.

Lemma 15.7. *Let Z be a set with* $\operatorname{card} Z = c$. *Then there are subsets Z_r, $r \in \mathbb{R}$, such that:*
(a) $\operatorname{card} Z_r = c$ *and* $Z = \bigcup_{r \in \mathbb{R}} Z_r$.
(b) $Z_r \cap Z_{r'} = \{z_{rr'}\}$ *is a singleton set for* $r \neq r'$.
(c) *If $r \neq r'$ or $s \neq s'$, then* $z_{rr'} \neq z_{ss'}$.
(d) *For each r, there is a $z_r \in Z_r$ such that* $z_r \notin \bigcup_{r' \neq r} Z_{r'}$.

Proof. We identify Z with the set $\{(a,b) \in \mathbb{R}^2 : 0 \leqslant a, 0 \leqslant b, a+b \leqslant 1\}$ in the plane. That is the triangle and its interior in Fig. 2.

For r in $\mathbb{R}$, let $T_r = \{(a,r) : 0 \leqslant a \leqslant 1-r\} \cup \{(1-r,b) : 0 \leqslant b \leqslant r\}$. Then $r_{r'r} = (1-r,r')$ if $r' < r$, and $z_r = (1-r,r)$. The remainder of the statements are clear from the figure. □

Now a large class of Dedekind domains is exhibited for which $(\operatorname{Prin} A)_+$ is dense in $\operatorname{Div} A$.

Proposition 15.8. *Let A be a Dedekind domain which contains a field k. If* $\operatorname{card} k \geqslant \operatorname{card} Z(A)$, *then $(\operatorname{Prin} A)_+$ is dense in $\operatorname{Div} A$.*

Proof. Let Y be a subset of $Z(A)$ with $\operatorname{card} Y < \operatorname{card} Z(A)$. As usual, for $\mathfrak{p}$ in $Z(A)$ pick f, g in A such that $\mathfrak{p} = (f, g)$ where $v_\mathfrak{p}(f) = 1$, $v_\mathfrak{p}(g) = 2$. Then $v_\mathfrak{p}(f + \lambda g) = 1$ for all λ in k. If each $f + \lambda g$ were contained in another prime ideal q in Y, there would be at least two distinct λ, λ' in k such that $f + \lambda g$ and $f + \lambda' g$ were in the prime ideal q. But then, as in the proof of Proposition 15.1, the ideal $q = p$. Hence, in fact, each

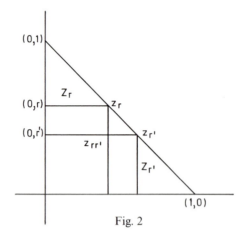

Fig. 2

prime ideal p in Y is principal with respect to Y. This certainly implies that $\rho(\mathrm{Prin}\, A)_+ = \mathbb{Z}_+^{(Y)}$. ☐

For the next proposition and its corollary, we assume the continuum hypothesis.

Proposition 15.9. *Let A be a Dedekind domain. If* $\mathrm{card}\, Z(A)$ *is not the successor of a limit cardinal, then* $\mathrm{Prin}\, A$ *is dense in* $\mathrm{Div}\, A$.

Proof. Suppose the subset Y of $Z(A)$ is a subset for which some element is not principal with respect to it. Then $\mathrm{card}\, A \leqslant (\mathrm{card}\, Y)^{\aleph_0}$ by Theorem 13.8. Hence, by Lemma 15.2, $\mathrm{card}\, Z(A) \leqslant (\mathrm{card}\, Y)^{\aleph_0}$. If $\mathrm{card}\, Y$ is not a limit cardinal, then $(\mathrm{card}\, Y)^{\aleph_0} = \mathrm{card}\, Y$ and hence $\mathrm{card}\, Y = \mathrm{card}\, Z(A)$. If $\mathrm{card}\, Y$ is a limit cardinal, then $\mathrm{card}\, Y < \mathrm{card}\, Z(A) = (\mathrm{card}\, Y)^{\aleph_0}$ and $\mathrm{card}\, Z(A)$ is its successor. ☐

Corollary 15.10. *If* $\mathrm{card}\, Z(A)$ *is not the successor of a limit cardinal, if* $\mathrm{card}\, A$ *is not the successor of* $\mathrm{card}\, Z(A)$, *and if* $\mathrm{card}\, Z(A) < \mathrm{card}\, A$, *then A is factorial.* ☐

Next we state and prove the crucial lemma and its corollaries, which are used for the construction of both Dedekind domains which realize pairs and Dedekind domains whose principal divisors are not dense.

Lemma 15.11. *Let A be a Dedekind domain which contains an algebraically closed field, k, of characteristic zero. Suppose the residue class fields are all isomorphic to k. Let K be the field of quotients of A.*

Suppose either that A is countable or $\text{trans} \cdot \deg_k K = 1$. *Suppose further that $Z(A)$ is infinite.*

Let $p(X)$ in $A[X]$ be a polynomial which is irreducible in $K[X]$ and let n be a non-negative integer. Then, for almost all $\mathfrak{p}$ in $Z(A)$, there is a maximal ideal $\mathfrak{m}$ in $A[X]$ such that

(1) *$\mathfrak{m}$ is the center on $A[X]$ of a discrete rank one valuation v defined on $K(X)$,*

(2) *the valuation v extends $v_{\mathfrak{p}}$ and the valuation ring V contains $A[X]$,*

(3) *the value $v(p(X)) = n$, and*

(4) *the residue class field of v is isomorphic with k.*

Proof. Suppose that $n > 0$; the case $n = 0$ being handled later. Choose a prime ideal $\mathfrak{p}$ in $Z(A)$ which contains no nonzero coefficient of $p(X)$ and which is also relatively prime to its discriminant. Then, in the ring $(A/\mathfrak{p})[X] (\cong k[X])$, $\overline{p}(X)$ factors into distinct linear factors; $\overline{p}(X) = \alpha_0 (X - \alpha_1) \cdots (X - \alpha_m)$ with each α_j in k. Let $g_1(X) = X - \alpha_1$ and $g_2(X) = \alpha_0 (X - \alpha_2) \cdots (X - \alpha_m)$. Then $g_1(X)$ and $g_2(X)$ are relatively prime in $(A/\mathfrak{p})[X]$. By Hensel's Lemma, there are a linear monic polynomial $G_1(X)$ and $G_2(X)$ in $\hat{A}_{\mathfrak{p}}[X]$ ($\hat{A}_{\mathfrak{p}}$ is the $\mathfrak{p}$-adic completion of $A_{\mathfrak{p}}$) such that $p(X) = G_1(X) G_2(X)$ in $\hat{A}_{\mathfrak{p}}[X]$, $\deg G_2 = \deg g_2(X)$, $\overline{G}_1(X) = g_1(X)$ and $\overline{G}_2(X) = g_2(X)$ (where $^-$ denotes reduction modulo $\mathfrak{p}$). Hensel's Lemma applies here, even though $p(X)$ is not necessarily monic, because its leading coefficient is a unit in $\hat{A}_{\mathfrak{p}}$.

Let $H_1(X)$, monic and linear, and $H_2(X)$ be polynomials in $A[X]$ which, modulo $\mathfrak{p}^{n+1}$, approximate $G_1(X)$ and $G_2(X)$ respectively. That is: $H_i(X) \equiv G_i(X) \pmod{\mathfrak{p}^{n+1} \hat{A}_{\mathfrak{p}}[X]}$ for $i = 1, 2$. Since either $\text{card } \hat{A}_{\mathfrak{p}} = 2^{\aleph_0}$ if A is countable or $\text{trans} \cdot \deg_k \hat{K}_{\mathfrak{p}} \geq 2$ in the other case [Zariski and Samuel, 1960, p. 220], there is a t in $\hat{A}_{\mathfrak{p}}$, transcendental over K, with $v_{\mathfrak{p}}(t) = n$ (where $v_{\mathfrak{p}}$ is the extension of the $\mathfrak{p}$-adic valuation on A to $\hat{A}_{\mathfrak{p}}$). Since t is transcendental over K and since $K[X] \cong K[H_1(X)]$, the map $\sigma: K[X] \to \hat{K}_{\mathfrak{p}}$, defined by extending the assignment $H_1(X) \mapsto t$, is an injection defined over K. Hence $K(X) \cong K(t) \subseteq \hat{K}_{\mathfrak{p}}$.

Define $v: K(X)^* \to \mathbb{Z}$ by composing σ with $v_{\mathfrak{p}}$; that is, $v = v_{\mathfrak{p}} \circ \sigma$. Since σ is defined over K, v extends $v_{\mathfrak{p}}/A$. Also $\sigma(A[X]) \subseteq \hat{A}_{\mathfrak{p}}$, so $v(A[X]) \subseteq \mathbb{Z}_+$. Thus the valuation ring V of v contains $A[X]$. The residue class field of v, being the same as that of $v_{\mathfrak{p}}$, is k. Let $\mathfrak{m}$ be the center of v on $A[X]$. Obviously $\mathfrak{p}A[X] \subseteq \mathfrak{m}$. If we can show that $p(X) \in \mathfrak{m}$, then it will be clear that $\mathfrak{m}$ is a maximal ideal in $A[X]$. For $p(X) \notin \mathfrak{p}A[X]$, so $\text{ht } \mathfrak{m} = \text{ht } \mathfrak{p}A[X] + 1$. But $\dim A[X] = 2$ and hence $\mathfrak{m}$ is a maximal ideal. We show that $p(X) \in \mathfrak{m}$ by showing that $v(p(X)) = n$, which will at the same time verify the remaining condition.

Since $p(X) - H_1(X) H_2(X)$ is in $\mathfrak{p}^{n+1} A[X]$, it has valuation greater than n. Also $v(H_2(X)) = 0$. Hence $n = v_{\mathfrak{p}}(t) = v(H_1(X)) = v(H_1(X) \cdot H_2(X))$.

By the triangle inequality for the discrete valuation v we have $v(p(X)) = v(H_1(X)) = n$.

Now we take the case $n=0$. Pick the same p as above and α in k such that $X - \alpha$ is relatively prime to $p(X)$ in $(A/p)[X]$. By the above, there are infinitely many prime ideals q in $Z(A)$ satisfying the conditions of the lemma for the polynomial $X - \alpha$ and the integer 1. Since $A[X]$ is a Hilbert ring, there are infinitely many of these prime ideals whose maximal ideals do not contain $p(X)$, but which satisfy all the other conditions. Hence the associated valuations are zero on $p(X)$. $\square$

The corollaries which follow will be used later.

Corollary 15.12. *Let $f(X, Y)$ be irreducible in $k[X, Y]$ and n an integer $n \geqslant 0$. Let Z be the set of maximal ideals in $k[X, Y]$, each which is a center on $k[X, Y]$ of a valuation with residue field k, valuation ring containing $k[X, Y]$ and value n at f. Then* $\operatorname{card} Y = \operatorname{card} k$.

Proof. If $f \notin k[Y]$ apply the lemma to $A = k[Y]$. There are $\operatorname{card} k$ prime ideals in A and only finitely many of these are excluded. $\square$

Corollary 15.13. *Given any set $T \subseteq k[X, Y]$ such that $\operatorname{card} T < \operatorname{card} k$, then there are $\operatorname{card} k$ valuations whose centers on $k[X, Y]$ are not zeros of any element in T. In other words, if $V(t) = \{ m \in Z(A); t \in m \}$, then* $\operatorname{card} \bigcap_{t \in T} (Z(A) - V(t)) = \operatorname{card} k$. *(We have set $A = k[X, Y]$.)*

Proof. The curves in A_k^2 defined by the elements of T cut any line $X = \alpha$ or $Y = \beta$ in at most $\operatorname{card} T$ points unless $X - \alpha$ or $Y - \beta$ is a component of one of the curves. There are at most $\operatorname{card} T$ components, leaving $\operatorname{card} k$ lines. On each of these lines there are $\operatorname{card} k$ centers available. $\square$

The next result is closely related to Lemma 15.11.

Proposition 15.14. *Let k be an algebraically closed field and X and Y indeterminates. Suppose $q(X, Y)$ is an irreducible polynomial. Let $\{ v_s \}_{s \in S}$ be a collection of valuations on $k(X, Y)$ with distinct centers satisfying the requirements:*
(a) Their valuation rings contain $k[X, Y]$.
(b) Their residue class fields are each k.
(c) $\operatorname{card} S < \operatorname{card} k$.
(d) The value $v_s(q(X, Y)) = n$ for all s.
Let f be in $\mathbb{Z}_+^{(S)}$. Then there is an irreducible polynomial $p(X, Y)$ in $k[X, Y]$ such that $v_s(p) = f(s)$ for all s in S.

Proof. Let m be one of the centers with valuation v. Then m is generated by $\{ X - a, Y - b \}$ for a and b in k. Either $v(X - a) = 1$ or $v(Y - b) = 1$ according to the construction of v. Suppose $v(Y - b) = 1$.

Let c be in k and set $u_c = (X-a) + c(Y-b)$. Suppose c and c' in k are distinct elements such that $v(u_c) > 1$ and $v(u_{c'}) > 1$. Then $v(Y-b) = v(u_c - u_{c'}) > 1$, a contradiction. Hence $v\big((X-a) + c(Y-b)\big) > 1$ for at most one c in k. Hence there is a c in k such that $v(u_c) = 1$ and $w(u_c) = 0$ for $w \neq v$ and $w \in \{v_s\}_{s \in S}$. Call this element h_s (where $v = v_s$). Let $g(X, Y) = \prod_{s \in S} h_s^{f(s)}$. Then g has the property: $v_s(g) = f(s)$ for all s in S.

For each s for which $f(s) > 0$, choose a linear polynomial h'_s (a line $h'_s = 0$) with $h'_s \neq h_s$, and $v_t(h'_s) = 0$ for $s \neq t$ and $v_s(h'_s) > 0$. Choose integers $n_s > f(s)$ such that the pencils of curves $p_d = g(X, Y) + d \prod_{f(s)>0} (h'_s)^{n_s} = 0$, $d \in k$, is not composed of a pencil. Then almost all of the elements in the pencil are irreducible [Zariski, 1941]. The value of any element in the pencil at v_s is clearly $f(s)$ when $f(s) > 0$. If p_d and $p_{d'}$, $d \neq d'$, both pass through another center where $f(s) = 0$, then $p_d - p_{d'} = (d - d') \prod (h'_s)^{n_s}$ also passes through this center, a contradiction to the choice of the h'_s. Hence $p(X, Y) = p_d(X, Y)$ where p_d is irreducible satisfies the requirement. □

The next result is used in constructing Dedekind domains, both in order to show that a free group with dense subset can be realized but, more importantly, to show that the principal divisors need not always be dense.

Proposition 15.15. *Let A be a countable Dedekind domain with field of quotients K. Suppose A contains an algebraically closed field k of characteristic 0 as a subring, that $Z(A)$ is infinite, and that A has k as residue class fields. Let X be an indeterminate. There is a Dedekind domain A' satisfying the following conditions:*

(1) *The field $K(X) \supset A' \supset A[X, X^{-1}]$.*

(2) *The induced function $\operatorname{Spec} A' \to \operatorname{Spec} A$ is a bijection.*

(3) *Each residue class field is isomorphic to k.*

(4) *Given any $q(X)$ in $K(X)$, there is an element g in A such that $g^{-1} q(X)$ is a unit in A'.*

Proof. Order the set of prime ideals $Z(A)$ and a set of representatives of the polynomials in $A[X]$ which are irreducible and pairwise relatively prime in $K[X]$ beginning with $p_1(X) = X$, $p_2(X)$, $p_3, \dots$. Let $\mathfrak{p}_1, \dots, \mathfrak{p}_k$ be the prime ideals of A containing the content of $p_2(X)$. By Lemma 15.11 above, produce valuations $v_1, \dots, v_k$ of $K(X)$ satisfying the conditions of that lemma, such that $v_j(X) = 0$ for each $j = 1, \dots, k$, and such that v_j extends $v_{\mathfrak{p}_j}$. There is an f in A such that $v_{\mathfrak{p}_j}(f) = v_j(p_2(X))$ for each j by the approximation theorem.

Again use Lemma 15.11 to produce valuations $v_{k+1}, \dots, v_{k+s}$ extending $v_{q_1}, \dots, v_{q_s}$, where these are the prime ideals containing f, such that they each vanish at X but with $v_{k+i}(p_2(X)) = v_{q_i}(f)$.

Let $r_1, \dots, r_t$ be the prime ideals in A not among the prime ideals $p_1, \dots, p_k, q_1, \dots, q_s$, which contain the content of $p_3(X)$ together with the first prime ideal in the ordering not taken previously. Produce valuations $v_{k+s+1}, \dots, v_{k+s+t}$, according to Lemma 15.11, which extend, respectively, the valuations associated to $r_1, \dots, r_t$ and vanishing on both X and $p_2(X)$. By the approximation theorem, there is an element b in A such that $v_i(b) = v_i(p_3(X))$ for $1 \leq i \leq k+s+t$. Continue as before.

We then get a family of valuation rings $\{V_i\}$, one for each prime ideal in A, whose maximal ideals lie over the corresponding prime ideal in A and each of which satisfies conditions (1), (3) and (4) of the statement of the proposition. The intersection $A' = \bigcap V$ is then the required Dedekind domain.   □

Example 15.16. We now construct a Dedekind domain A' whose principal divisors are not dense in $\text{Div}\, A'$. We want a Dedekind domain A' and a subset $Y \subset Z(A')$ such that $\text{card}\, Y < \text{card}\, Z(A')$ and such that not every prime in Y is principal with respect to Y. It is enough to find a Y such that the subintersection $\bigcap_{p \in Y} A'_p = A'_Y$ is not factorial. For if every prime ideal in Y is principal with respect to Y, then, indeed A'_Y must be factorial.

We need a countable Dedekind domain A containing an algebraically closed field k of characteristic 0 with residue class fields isomorphic to k which is not factorial. For example, $k[U, Y]/(U^3 - U - Y^2) = A$, where k is the algebraic closure of $\mathbb{Q}$ in $\mathbb{C}$. Let $X = \{X_t\}_{t \in [0, 1]}$ be a family of indeterminates.

For subsets $Y \subset X$, let B_Y be a Dedekind domain such that
(1) $K(X) \supset K(Y) \supset B_Y \supset A[Y, Y^{-1}]$ (where $K(X) = K(\{X_t\}_{t \in [0, 1]})$, etc.),
(2) $\text{Spec}\, B_Y \to \text{Spec}\, A$ is a bijection,
(3) The residue fields of B_Y are k, and
(4) Given $q \neq 0$, $q \in K(Y)$, there is an f in A such that $f^{-1}q$ is a unit in B_Y.

Consider the family of all such pairs (Y, B_Y) ordered by inclusion. By Proposition 15.15, this family is not empty. Any chain in the family has an upper bound, the union, as is easily verified. Hence there is a maximal pair (M, B_M). If M, and hence B_M, were countable, there would be an $X_\alpha \notin M$, $0 \leq \alpha \leq 1$, and we could extend further, by Proposition 15.15. Thus any maximal element (M, B_M) must have $\text{card}\, M = \text{card}\, X$ and $\text{card}\, B_M = \text{card}\, X$. But $\text{Spec}\, B_M \cong \text{Spec}\, A \cong Z(A)$ is countable and B_M is not factorial $(A \to B_M$ induces a bijection $\text{Cl}(A) \to \text{Cl}(B_M))$.

Now apply Lemma 15.7 in order to subdivide the set M into subsets satisfying the requirements stated there. If L and N are subsets, let X_{LN} denote the variable in $L \cap N$ and X_{NL} the variable in $N \cap L$. Let X_L

denote the unique element in L not in any other subset. For each subset L choose a prime ideal $\mathfrak{p}_L$ in $K[M]$ by setting $\mathfrak{p}_L = (X_L, \{X_\sigma - \alpha_\sigma\}_{X_\sigma \in L - \{X_L\}})$ where $\alpha_\sigma \in k$ subject only to the condition that $\alpha_{LN} \neq \alpha_{NL}$ for each pair (L, N). Let v_L denote the $\mathfrak{p}_L$-adic valuation of $K[M]$ with associated valuation ring V_L.

Let $A' = B_M \cap \left(\bigcap_L V_L \right)$. We wish to show that A' is a Dedekind domain. For this it is necessary to verify the finite character property, so that A' is a Krull domain, and to show that the Krull dimension is one, or that the nonzero prime ideals are maximal. To see the finite character property, it is sufficient to show that every element in $A[M]$ is a unit in almost all the valuation rings. Let f be in $A[M]$. Since B_M is a Dedekind domain, f lies in at most finitely many prime ideals in B_M. Furthermore f is defined by a finite number of indeterminates and so it lies in at most a finite number of primes $\mathfrak{p}_L$. Thus f is a unit in the other $V_{L'}$. To see the maximality of the nonzero prime ideals, it is sufficient to show that two distinct minimal nonzero prime ideals in A' generate A'. If these two distinct divisorial prime ideals arise from B_M, their centers on A are relatively prime. Therefore the sum of their centers on A, hence the sum of the ideals themselves, contains 1. If the two distinct divisorial prime ideals are centers on A' of V_L and $V_{L'}$, then their respective centers on $A[M]$ contain $\mathfrak{p}_L$ and $\mathfrak{p}_{L'}$. But in $\mathfrak{p}_L + \mathfrak{p}_{L'}$ lies the element $\alpha_{LN} - \alpha_{NL} = (X_{NL} - \alpha_{NL}) - (X_{LN} - \alpha_{LN})$ which is invertible, as it lies in k. As for the third possibility, the centers in $A[M]$ of a prime ideal in B_M and a V_L, one has X_L in the ideal defined by V_L while $X_L - \alpha$ is in the ideal coming from B_M for some $\alpha \in k$, $\alpha \neq 0$. Hence these two centers together generate $A[M]$.

Now A' is a Dedekind domain with $Z(A') = Z(B_M) \cup M$, so $\operatorname{card} Z(A') = \operatorname{card} M$, while $\operatorname{card} Z(B_M) = \aleph_0$. Hence B_M is a subintersection which is not factorial. Thus $\operatorname{Prin} A'$ is not dense in $\operatorname{Div} A'$. $\square$

On the other hand, it is possible to construct Dedekind domains which realize pairs (X, P).

Recall that a cardinal α is *regular* if α cannot be written as a sum $\sum_{n \in W} \gamma_n$, where $\operatorname{card} W < \alpha$ and $\gamma_n < \alpha$ for all $n \in W$. If α is not a limit cardinal, it is regular. On the other hand, if α is not regular, it can be written as the sum of non-limit (and hence regular) cardinals. This being said, we assume first that $\operatorname{card} X = \alpha$ is a regular cardinal.

Let k be an algebraically closed field of characteristic 0 with $\operatorname{card} k = \alpha$.

Lemma 15.17. Let $\{ f_s(U, Y) \}_{s \in S}$ be a family of elements in $k[U, Y]$ where $\operatorname{card} S < \operatorname{card} k$. Let p be an irreducible polynomial in $k[U, Y]$. Let X be a set with $\operatorname{card} X = \operatorname{card} k$ and let f be in $\mathbb{Z}^{(X)}$. Then there is *a family of valuations* $\{v_x\}_{x \in X}$ *whose valuation rings contain* $k[U, Y]$, *which have distinct centers which are maximal ideals in* $k[U, Y]$ *and which satisfy, in addition:*

    (i) *For all* x *in* X, $v_x(p) = f(x)$.

    (ii) *For all* x *in* X *and all* s *in* S, $v_x(f_s) = 0$.

    (iii) *If* g *is irreducible and* $\deg g = n$, *then* $\operatorname{card}\{x : v_x(g) \neq 0\} \leqslant n^2 + 1$.

Proof. Well order the elements in X and go by transfinite induction. Suppose, for some x in X, we have produced valuations $v_{x'}$ for all $x' < x$ satisfying the requirements. For each finite subset C of the centers $\{\mathfrak{m}_{x'}\}_{x' < x}$ choose a polynomial p_C of smallest degree lying in each center in C. Consider the set $W = \{p_C : C\} \cup \{f_s\}_{s \in S}$. By Corollary 15.14, choose a maximal ideal $\mathfrak{m}_x$ which is a center of a valuation v_x such that $v_x(p) = f(x)$, distinct from the centers $\mathfrak{m}_{x'}$ and which contains no element in W. Now all the statements of the proposition are satisfied by the set of valuations $\{v_{x'}\}_{x' \leqslant x}$ except possibly (iii). Let g be an irreducible polynomial of degree n. If g is in $n^2 + 2$ of the centers, then g must lie in $\mathfrak{m}_x$, by the induction hypothesis, and $n^2 + 1$ of the other centers. A curve of smallest degree passing through these other centers has degree at most n ($g = 0$ is one) and degree at least n by hypothesis. But then it can have at most n^2 centers in common with $g = 0$ unless it is associated to g, by Bezout's theorem. This is a contradiction.   ☐

We are now prepared to prove the main result, the realization theorem.

Theorem 15.18. *Let* X *be a set and* P, *contained in* $\mathbb{Z}_+^{(X)}$, *a dense subset. Then there is a Dedekind domain* A *and a bijection* $\sigma : Z(A) \to X$ *such that the induced isomorphism* $\sigma : \mathbb{Z}^{(X)} \to \operatorname{Div} A$ *induces an isomorphism* $\sigma : \operatorname{Gr} P \to \operatorname{Prin} A$, *where* $\operatorname{Gr} P$ *denotes the group generated by* P.

Proof. We assume first that $\operatorname{card} X$ is regular. Well order, to minimal order type, the elements of X, and a set of representatives of the irreducible elements of $k[U, Y]$ (where k is algebraically closed, characteristic zero with $\operatorname{card} k = \operatorname{card} X$). Call these the lists.

For each x in X, let P_x denote the subgroup of $\mathbb{Z}^{(X)}$ generated by those f in P such that $f(x') = 0$ for all $x' > x$.

Consider triples $D_x = (x, \{v_{x'}\}_{x' \leqslant x}, S_x)$ where $x \in X$, $\{v_{x'}\}_{x' \leqslant x}$ is a family of valuations whose valuation rings contain $k[U, Y]$ and whose centers are distinct maximal ideals, and S_x is a family of irreducible elements in the list. These three satisfy the conditions:

    (i) If $g(U, Y)$ is irreducible of degree n, then it lies in at most $n^2 + 1$ centers of the $\{v_{x'}\}$.

    (ii) The functions f defined by $f(x') = v_{x'}(h)$ for $x' \leqslant x$ and h in S_x generate P_x.

    (iii) The cardinality $\operatorname{card} S_x \leqslant \operatorname{card} [1, x]$.

Define a relation "$D_{x'} < D_x$" if (a) $x' < x$, (b) on the interval $[1, x']$ the valuations of $D_{x'}$ coincide with the valuations of D_x, (c) card $[x', x]$ is infinite and (d) S_x contains the first element in the list of irreducibles which lies outside of $S_{x'}$. Let $\leqslant$ denote the associated partial order.

Select a maximal chain of these triples $\{D_x\}$. The x which occur in this chain are not bounded in X. For otherwise, suppose $x \leqslant x'$. Choose the first irreducible t in the list which is outside $\bigcup_x S_x$. $\left(\text{card} \bigcup_x S_x\right.$ $\leqslant \text{card}[1, x'] < \text{card}(\text{list}).\big)$ Choose f in P which has for its coefficient on x'' the integer $v_{x''}(t)$ for any valuation arising from the triples D_x. This is possible since P is dense. Choose f such that $f(x'') > 0$ for some x'' where $[x', x'']$ is infinite. Now for any $y < x''$ such that $x \leqslant y$ for all D_x in the chain, choose a valuation v_y such that $v_y(t) = f(y)$. Choose a set T of irreducible elements in the list with card $T = \text{card}[1, y]$ whose values generate P_y. The new triple D_y with $S_y = \left(\bigcup_x S_x\right) \cup T$ properly extends the chain, contrary to the assumption that the chain was maximal.

Hence, for each x in X, there is a valuation v_x with valuation ring V_x. Let A be the intersection $\bigcap_x V_x$. Then A is a Dedekind domain. Every element in P is realized by the function associated to some irreducible element in $k[U, Y]$ and $\text{Div } A = \mathbb{Z}^{(X)}$ is induced by $Z(A) \cong X$. It remains to show that every irreducible element defines a function in the subgroup generated by P. Here the maximality of the chain is used.

Let p_x be the first irreducible not in S_x. The function defined by $x \mapsto p_x$ is increasing. There must be card k distinct x (card k is regular) and the list is ordered to minimal order type. Hence every irreducible is in some S_x.

There remains the problem of non-regular card X. One more lemma is needed.

Lemma 15.19. *Let P be a dense subset of $\mathbb{Z}_+^{(X)}$. Let X' be a subset of X where* card X' *is not a limit cardinal. Then there is a subset* $X'' \subset X$ *such that* card $X'' = $ card X' *and the set* $P_{X''} = \{f \in P : f(x) = 0 \, \forall x \notin X''\}$ *is dense in* $\mathbb{Z}^{(X'')}$.

Proof. Let $Y \subset X'$ be such that card $Y < $ card X'. Let $y \in Y$. There is an element f_y in P such that $f_y(y) = 1$ and $f_y(y') = 0$ for all y' in Y, with $y' \neq y$. Let $(X')_{y, Y} = X' \cup \{x \in X : f_y(x) > 0\}$. Then define $(X')_1$ $= \bigcup_{y \in Y, Y \subset X'} (X')_{y, Y}$. Choose a set Γ with card $\Gamma = $ card X such that Γ is well ordered to minimal order type. For each γ in Γ define a subset $X'_\gamma \subset X$ by (1) $X'_\gamma = (X'_{\gamma - 1})_1$ if γ is not a limit ordinal and (2) $X' = \bigcup_{\beta < \gamma} X'_\beta$ otherwise. Let $X'' = \bigcup_{\gamma \in \Gamma} X'_\gamma$.

It follows, by transfinite induction, that $\operatorname{card} X'_\alpha = \operatorname{card} X'$ for all α in Γ. Hence $\operatorname{card} X'' = \operatorname{card} X'$. If $U \subset X''$ with $\operatorname{card} U < \operatorname{card} X''$, then $U \subset X'_\alpha$ for some α in Γ. Hence $X'_{\alpha+1}$ contains a function in P yielding arbitrary coefficients on U and 0 outside $X'_{\alpha+1}$. $\square$

Now let X and P be given, where $\operatorname{card} X$ is not regular. Write $X = \bigcup X_t$ where $\operatorname{card} X_t$ is not a limit cardinal and P_t is dense in X_t. Write k as the union of algebraically closed subfields k_t where $\operatorname{card} k_t = \operatorname{card} X_t$. Each X_t has a regular cardinal, so we can find a Dedekind domain A_t with $k_t[U, Y] \subseteq A_t \subseteq k_t(U, Y)$ which realizes (X_t, P_t). If $X_t \supset X_{t'}$, we can make certain that $A_{t'} \subset A_t$ in such a way that the valuations of A_t restrict to those of $A_{t'}$. Then $A = \bigcup A_t$ is a Dedekind domain which realizes (X, P). $\square$

We now apply this construction to get some rather pathological Dedekind domains. The full generality is not needed here, only the countable case, which is much easier (cf. [Claborn, 1968]).

Example 15.20. *There is a Dedekind domain A with $\operatorname{Cl}(A)$ cyclic of order n, n a fixed integer, $n > 1$, such that all the prime ideals are in one class. I.e., if $\mathfrak{p}$ and $\mathfrak{q}$ are prime ideals, there is an x in K such that $x\mathfrak{p} = \mathfrak{q}$.*

Proof. Fix the integer n. Let $X = \mathbb{N} = \{1, 2, \ldots\}$. Let $P = \left\{ f \in \mathbb{Z}_+^{(\mathbb{N})} : \sum_i f(i) \text{ is divisible by } n \right\}$. The set P is clearly dense in X.

Let A realize $(\mathbb{N}, P)$. Then $\operatorname{Cl}(A) = \mathbb{Z}^{(\mathbb{N})}/\operatorname{Gr}(P) \cong \mathbb{Z}/n\mathbb{Z}$. If $\mathfrak{p}, \mathfrak{q}$ are prime ideals, then $\mathfrak{p}^n$ and $\mathfrak{p}^{n-1}\mathfrak{q}$ are principal, since the functions $n\operatorname{div}(\mathfrak{p})$ and $(n-1)\operatorname{div}\mathfrak{p} + \operatorname{div}(\mathfrak{q})$ are in $\operatorname{Gr}(P)$. $\square$

Example 15.21. *There is a Dedekind domain A with $\operatorname{Cl}(A) = \mathbb{Z}$ such that for any two prime ideals $\mathfrak{p}, \mathfrak{q}$, either $\mathfrak{p}\mathfrak{q}$ or $\mathfrak{p}\mathfrak{q}^{-1}$ is principal. So if c is a class which generates $\operatorname{Cl}(A)$, then either $\operatorname{cl}(\mathfrak{p}) = c$ or $\operatorname{cl}(\mathfrak{p}) = -c$.*

Proof. Let $P = \left\{ f \in \mathbb{Z}_+^{(\mathbb{N})} : \sum_i (-1)^i f(i) = 0 \right\}$. If we denote the coordinate function by x_i, then $P = \operatorname{Gr}(\{x_i + x_{i+1}\}_{i \geqslant 1})_+$. The set P is clearly dense and $\mathbb{Z}^{(\mathbb{N})}/\operatorname{Gr}(P) \cong \mathbb{Z}$. Further, the image of any coordinate function is a generator. Let A realize $(\mathbb{N}, P)$. If $\mathfrak{p}_i$ is the prime ideal corresponding to x_i, then $\mathfrak{p}_i \mathfrak{p}_{i+1}$ is principal for all i. Hence $\operatorname{cl}(\mathfrak{p}_i) + \operatorname{cl}(\mathfrak{p}_{i+1}) = 0$. The rest is easy to verify. $\square$

Remark. Let A be a Dedekind domain constructed in either of the two examples above. If A' is a ring properly between A and its quotient field K, then A' is factorial. For A' is still a Dedekind domain (Proposition 13.2) and is a subintersection. So $\operatorname{Cl}(A) \to \operatorname{Cl}(A')$ is a surjection. Its kernel is generated by the classes of the prime ideals omitted in forming the subintersection (Nagata's Theorem 7.1). But the class of any prime ideal generates $Cl(A)$. Hence the map is zero.

Thus a Dedekind domain can be locally factorial (i.e., A_f factorial for all non-units f in A) but not factorial.

Problem. Is there a local (noetherian) Krull domain A such that A_f is factorial for all non-units f which is not factorial? Note that if A is the local ring $k[x,y,u,v]_{(x,y,u,v)}$ where $xv=yu$, then A_w is factorial where w runs through the set $\{x,y,u,v\}$ of generators for the maximal ideal. Thus $\operatorname{Spec} A - \{m\}$, the punctured spectrum, is locally factorial, but A is not factorial.

Example 15.22. *There is a Dedekind domain A such that no semi-prime ideal is principal.*

A semi-prime ideal is an ideal which is the product of distinct prime ideals.

Suppose that D is a principal ideal domain, that L is a finite separable extension of the field of quotients of D, and that C is a subring of L whose field of quotients is L and which contains the integral closure of D. Then C contains a principal semi-prime ideal. Thus the example A constructed above is a Dedekind domain which cannot be the subintersection of a finite extension of a principal ideal domain. (Confer also: [Zariski-Samuel, 1958], [Claborn, 1965c], [Leedham-Green, 1972].)

Proof. Let $P = \left\{ f \in \mathbb{Z}_+^{(\mathbb{N})} : f = \sum_i n_i(x_i + 2\,x_{i+1}) \right\}$. Let A realize $(\mathbb{N}, P)$. Since the last coefficient of every element in P is even, no principal ideal can be semi-prime. $\square$

There are further results in this direction, which we only mention here. Claborn [1968] is able to control, to some extent, the group of units of the integral domain which realizes a pair. In the countable case A can be chosen so that it has only $\{\pm 1\}$ as a group of units.

Leedham-Green [1972] has considered related problems. He constructs an example of a Dedekind domain with no semi-prime principal ideals. He also is able to show, using essentially the same construction as reported above in Proposition 15.15, that a Dedekind domain with specified class group can be obtained as a quadratic extension of a principal ideal domain.

Chapter IV. Descent

Two descent techniques are the topics in this chapter; Galois and radical descent. As an application of these techniques we are able to construct examples of factorial rings with certain properties or we are able to construct non-factorial rings. A particular instance of this is M.-J. Bertin's [1967] example of a factorial ring which is not Cohen-Macaulay. Another instance is the construction of a factorial ring A whose formal power series ring is not factorial.

The basic material was presented first in [Samuel, 1964 b]. Recently Waterhouse [1971] has combined the two theories into one by the use of Hopf algebras and module algebras. The calculations he makes are similar to those made by Samuel. Thus a good understanding of Waterhouse's work necessitates reading Samuel's papers. For this reason the descent methods of Samuel are included. The interested reader is referred to Waterhouse's very good paper.

§ 16. Galois Descent

Suppose G is a finite group of automorphisms acting on the Krull domain B. Let A denote the fixed ring of G. The group acts also on the field of quotients L of B and the fixed subring K of L is the field of quotients of A. Since $A = B \cap K$, the ring A is also a Krull domain. Since G is a finite group, the ring B is integral over A. Hence the inclusion $A \rightarrow B$ induces a homomorphism $i : \mathrm{Cl}(A) \rightarrow \mathrm{Cl}(B)$. The problem is to study the kernel of this homomorphism.

Theorem 16.1. *There is induced a natural injection* $\theta : \ker i \rightarrow \mathrm{H}^1(G, B^*)$. *If every prime divisorial ideal of B is unramified over A, then θ is a bijection.*

Proof. We have exact sequences $1 \rightarrow B^* \rightarrow L^* \rightarrow \mathrm{Prin}\, B \rightarrow 0$ and $0 \rightarrow \mathrm{Prin}\, B \rightarrow \mathrm{Div}\, B \rightarrow \mathrm{Cl}(B) \rightarrow 0$.

Each element of G is an automorphism on B^* and L^*, so there is induced a G-module structure on $\mathrm{Prin}\, B$. The fixed subgroup functor yields the exact sequence of cohomology groups $1 \to B^{*G} \to L^{*G}$ $\to (\operatorname{Prin} B)^G \to H^1(G, B^*) \to H^1(G, L^*)$. But $H^1(G, L^*) = 0$ by Hilbert's Theorem 90. Also $B^{*G} = A^*$ and $L^{*G} = K^*$, so the sequence becomes: $1 \to A^* \to K^* \to (\operatorname{Prin} B)^G \to H^1(G, B^*) \to 0$. But G acts also on $\operatorname{Div} B$ and hence on $\operatorname{Cl}(B)$. We get the left exact sequence $0 \to (\operatorname{Prin} B)^G$ $\to (\operatorname{Div} B)^G \to \operatorname{Cl}(B)^G$. The image of $\operatorname{Div} A$ in $\operatorname{Div} B$ consists of G-invariant elements. Thus the induced homomorphism $i : \operatorname{Cl}(A) \to \operatorname{Cl}(B)$ factors through $\operatorname{Cl}(B)^G$. We get the commutative diagram of exact sequences

$$
\begin{array}{ccccccccc}
 & & & & 0 & & & & \\
 & & & & \downarrow & & & & \\
0 & \to & \operatorname{Prin} A & \to & \operatorname{Div} A & \to & \operatorname{Cl}(A) & \to & 0 \\
 & & \downarrow & & \downarrow & & \downarrow & & \\
0 & \to & (\operatorname{Prin} B)^G & \to & (\operatorname{Div} B)^G & \to & \operatorname{Cl}(B)^G & & \\
 & & \downarrow & & & & & & \\
 & & H^1(G, B^*) & & & & & & \\
 & & \downarrow & & & & & & \\
 & & 0 & & & & & &
\end{array}
$$

with exact rows. Clearly $\operatorname{Div} A \to \operatorname{Div} B$ is an injection, hence $\operatorname{Div} A \to (\operatorname{Div} B)^G$ is an injection. Therefore there is induced a unique homomorphism $\theta : \ker i \to H^1(G, B^*)$ which is an injection.

The homomorphism θ is a bijection if and only if $\operatorname{Div} A \to (\operatorname{Div} B)^G$ is a surjection. Suppose B is unramified over A at each divisorial prime ideal of B. Let $\mathfrak{P}$ be such a prime ideal. Then $\bigcap_{\sigma \in G} \sigma(\mathfrak{P})$ is a divisorial ideal of B which is the divisor of the extension of $\mathfrak{P} \cap A$ to B.

Suppose the divisor $d = \sum_{\mathfrak{P} \in X^{(1)}(B)} n_{\mathfrak{P}} \operatorname{div}(\mathfrak{P})$ is in $(\operatorname{Div} B)^G$. Now $\sigma(d) = \sum_{\mathfrak{P} \in X^{(1)}(B)} n_{\mathfrak{P}} \operatorname{div}(\sigma(\mathfrak{P}))$. We write $d = \sum_{\mathfrak{p} = \mathfrak{P} \cap A} \left(\sum_{\mathfrak{P}} n_{\mathfrak{P}} \operatorname{div}(\mathfrak{P}) \right)$. Then $d = \sum_{\mathfrak{p} = \mathfrak{P} \cap A} \left(\sum n_{\sigma(\mathfrak{P})} \operatorname{div}(\sigma(\mathfrak{P})) \right)$. Since $\sigma(d) = d$, the integers $n_{\sigma(\mathfrak{P})}$ are constant. Hence $d = \sum_{\mathfrak{p}} n_{\mathfrak{p}} \left(\sum \operatorname{div}(\sigma(\mathfrak{P})) \right)$ which is the image of the divisor $\sum_{\mathfrak{p} \in X^{(1)}(A)} n_{\mathfrak{p}} \operatorname{div}_A(\mathfrak{p})$. Therefore $\operatorname{Div} A \to (\operatorname{Div} B)^G$ is a surjection so θ is a bijection. □

That nonramification is essential in order to assert that θ is a bijection is verified by a very classical example. Let B be the Gaussian integers $\mathbb{Z}[i]$ and G the Galois group of $\mathbb{Q}(i)$. Then $B^G = \mathbb{Z}$ and the extension $A \to B$ is ramified at the prime containing 2. Both A and B are factorial, whereas $H^1(G, B^*) = \mathbb{Z}/2\mathbb{Z}$. So θ is not a bijection.

There is a fine test for ramification of B over A. Since L is Galois over K, it is separable over K. Hence there is a primitive element u

which can be chosen in B. Let $T:L\to K$ be the trace function. Then the induced bilinear form $b:L\times L\to K$ given by $b(x,y)=T(xy)$ is non-degenerate. The free A-module generated by the powers of u is a finitely generated A-submodule of B which generates L over K.

Proposition 16.2. *The ring B is a divisorial A-lattice in L.*

Proof. We have a finitely generated free A-submodule of B which is an A-lattice in L. Let F be this free A-lattice, so $F=A[u]$ where u is a primitive element in B. Now $F^*=\{x\in L: T(xF)\subseteq A\}$. Then F^* is a free A-module (isomorphic to $\mathrm{Hom}_A(F,A)$). Since B is integral over A, the trace $T(BF)\subseteq A$. Hence we have B pinched between F and F^*, so B is an A-lattice in L.

We want to show that $B=\bigcap\limits_{\mathfrak{p}\in X^{(1)}(A)} B_\mathfrak{p}$. But $B_\mathfrak{p}$ is the integral closure of $A_\mathfrak{p}$ in L. So $B_\mathfrak{p}$ has a finite number of maximal ideals and each is the extension to $B_\mathfrak{p}$ of a divisorial prime ideal of B. Hence $B_\mathfrak{p}=\bigcap\limits_{\mathfrak{P}\cap A=\mathfrak{p}} B_\mathfrak{P}$. As each $\mathfrak{P}$ has a non-zero intersection with A, each $B_\mathfrak{P}$ appears somewhere. So B is divisorial. $\square$

Now the complementary module and the Dedekind different can be defined in the usual way. The *complementary module* $\mathfrak{C}(B/A)$ is the divisorial fractionary ideal $\{x\in L: T(xB)\subseteq A\}$. It is clear that $\mathfrak{C}(B/A)\supseteq B$. Also $\mathfrak{C}(B/A)_\mathfrak{p}=\mathfrak{C}(B_\mathfrak{p}/A_\mathfrak{p})$ for all $\mathfrak{p}$ in $X^{(1)}(A)$ by Corollary 5.5c. Hence $\mathfrak{C}(B/A)$ is a divisorial fractionary B-ideal in L.

The *Dedekind different* $\mathfrak{D}(B/A)$ is the ideal inverse of $\mathfrak{C}(B/A)$. That is $\mathfrak{D}(B/A)=\{x\in L: x\mathfrak{C}(B/A)\subseteq B\}$. Also for each $\mathfrak{p}$ in $X^{(1)}(A)$ we have $\mathfrak{D}(B/A)_\mathfrak{p}=\mathfrak{D}(B_\mathfrak{p}/A_\mathfrak{p})$.

Proposition 16.3. *The Krull domain B is unramified over A at $\mathfrak{P}$ in $X^{(1)}(B)$ if and only if $\mathfrak{D}(B/A)\not\subseteq\mathfrak{P}$. In particular B is unramified over A at all divisorial prime ideals if and only if $\mathfrak{D}(B/A)=B$.*

Proof. The ring B is unramified over A at $\mathfrak{P}$ if and only if $\mathfrak{P}B_{\mathfrak{P}\cap A}$ is unramified over $A_{\mathfrak{P}\cap A}$. But this is the case if and only if $\mathfrak{D}(B_{\mathfrak{P}\cap A}/A_{\mathfrak{P}\cap A})$ $\not\subseteq\mathfrak{P}B_{\mathfrak{P}\cap A}$. (See, for example, [Serre, 1962], [Artin, 1959], [Auslander and Buchsbaum, 1959], [Deuring, 1968], or any book on algebraic number theory.)

The global statement follows. $\square$

Let $f(t)$ be a minimal polynomial for the primitive element u in B. Denote its derivative by $f'(t)$. Then $\mathfrak{C}(B/A)\subseteq f'(u)^{-1}B$ and hence $f'(u)\in\mathfrak{D}(B/A)$. This is the case for all such u. If, for example, we can find primitive elements $u_1,\ldots,u_s$ in B such that $f_1'(u_1),\ldots,f_s'(u_s)$ are not in any prime divisorial ideal, then B is unramified over A at each divisorial prime ideal of B. This will be the case in many of the examples.

The most simple case is encountered when G is cyclic. Then we can calculate $H^1(G, B^*)$ in terms of other invariants. Let σ generate the cyclic group G. For x in L^*, define $h(x) = \sigma(x)/x$. Then $h: L^* \to L^*$ is a group homomorphism. Each such x in L^* can be written as a fraction $x = b/a$ where $b \in B$ and $a \in A$. Then $h(x) = h(b)$. Therefore $h(L^*) = h(B - \{0\})$. Then $H^1(G, B^*) = B^* \cap h(L^*)/h(B^*)$. Let h_1 be the restriction of h to B^*. Denote by N and N_1 the norm on L^* and its restriction to B^* respectively $\left(N(x) = \prod_{\sigma \in G} \sigma(x) \right)$. Then $B^* \cap h(L^*) = \ker N_1$ and $h(B^*) = \operatorname{im} h_1$. That is $H^1(G, B^*)$ is the homology of the complex $B^* \xrightarrow{h_1} B^* \xrightarrow{N_1} A^*$.

We use these calculations for the next result. Let B be the polynomial ring $k[X_1, \ldots, X_r]$. Suppose G is a cyclic group whose order is a power of the characteristic exponent of k which acts as a group of k automorphisms of B.

Proposition 16.4. *The fixed ring B^G is factorial.*

Proof. Since B is factorial, the group $\operatorname{Cl}(B^G)$ is isomorphic to a subgroup of $H^1(G, B^*)$. But $B^* = k^*$ as a trivial G-module. Hence $H^1(G, B^*) = \operatorname{Hom}_{\mathbf{Z}}(G, k^*)$. But k^* has no torsion of the order of its characteristic exponent. Hence $H^1(G, B^*) = 0$.

(Or one may calculate the kernel of N_1. But $N_1(x) = x^{\operatorname{ord} G}$. Hence $\ker N_1 = 1$.) □

In general when G is a finite group of k-automorphisms of $k[X_1, \ldots, X_r]$, then $H^1(G, k^*) = \operatorname{Hom}_{\mathbf{Z}}(G, k^*)$ which is the subgroup of the group of card G-roots of unity of k^*.

Example 16.5. Let k be a field of characteristic exponent p. Suppose n is an integer prime to p and suppose k contains a primitive n^{th} root of unity, say w. Define a k-automorphism of $k[X_1, \ldots, X_r]$ $(r \geq 2)$ by extending to all of $k[X_1, \ldots, X_r]$ the assignments $X_i \mapsto w X_i$. The cyclic group generated is isomorphic to $\mathbf{Z}/n\mathbf{Z}$. The fixed subring A is the subalgebra generated by the monomials of degree n. Each indeterminate X_i is a primitive element whose minimal polynomial is $t^n - X_i^n$. The value of the derivative is $n X_i^{n-1}$. The ideal generated clearly is not contained in a prime divisorial ideal, so the extension is unramified at the prime divisorial ideals since $r \geq 2$. Hence $\operatorname{Cl}(A) = H^1(G, k^*)$ which is just $\mathbf{Z}/n\mathbf{Z}$.

Example 16.5a. This same technique can be used for computing the ideal class group of the homogeneous coordinate ring of projective real and complex projective space. Let G be the cyclic group of order 2 with generator σ. Let A_n be the ring $\mathbb{R}[x_0, \ldots, x_n]$ with $x_0^2 + \cdots + x_n^2 = 1$

and B_n the complexification, so $B_n = \mathbb{C} \otimes_\mathbb{R} A_n$. Then G acts on A_n and B_n as ring automorphisms by extending the assignments $\sigma(x_i) = -x_i$ for each i. The fixed subrings are the rings generated by the quadratic monomials. If one divides the n-sphere S^n by antipodal points, then these subrings are the rings of homogeneous polynomial functions on the quotient spaces. The group of units is the group of non-zero field elements in all cases except for B_1. The ring $B_1 = \mathbb{C}[x_0, x_1] \cong \mathbb{C}[x_0 + ix_1, (x_0 + ix_1)^{-1}]$ has its group of units identified with $\mathbb{C}^* \times \mathbb{Z}$ with $\sigma(c, n) = ((-1)^n c, n)$. However it is easy to determine that $B_1^G = \mathbb{C}[(x_0 + ix_1)^2, (x_0 + ix_1)^{-2}]$ which is factorial. As for the other cases, we have $\mathrm{Cl}(A_n^G) = \mathbb{Z}/2\mathbb{Z}$ for all $n \geqslant 1$ and $\mathrm{Cl}(B_n^G) = \mathbb{Z}/2\mathbb{Z}$ for $n \geqslant 2$.

M.-J. Bertin [1967] has made use of Galois descent in order to construct an example of a factorial noetherian ring which is not Cohen-Macaulay.

Let k be a field of characteristic $p, p \neq 0$. Let $B = k[X_1, \ldots, X_r]$ be the polynomial ring in r variables over k. Each element of the special linear group $\mathrm{SL}(r, k)$ induces a k-automorphism of B. This action is given by letting each element act on a vector space V of dimension r over k and then extending to the symmetric algebra of V over k, this algebra being exactly B. If $\sigma = (a_{ij})$ and is an element in $\mathrm{SL}(r, k)$, then the action of σ on the indeterminants is given by $\sigma(X_i) = \sum_{j=1}^{r} a_{ji} X_j$ for $i = 1, \ldots, x$. We would like to take an element whose order is a power of the prime p. Then the fixed subring will be factorial. According to [Bertin, 1967] it is sufficient to study the fixed rings for the element

$$\begin{pmatrix} 1 & 0 & 0 & \ldots & 0 & 0 \\ 1 & 1 & 0 & \ldots & 0 & 0 \\ 0 & 1 & 1 & \ldots & 0 & 0 \\ \vdots & \vdots & \vdots & & \vdots & \vdots \\ 0 & 0 & 0 & \ldots & 1 & 0 \\ 0 & 0 & 0 & \ldots & 1 & 1 \end{pmatrix}$$

in $\mathrm{SL}(r, k)$.

For this σ the ring $k[X_1, \ldots, X_r]$ is unramified over the fixed ring at the divisorial prime ideals. However since the cohomology group is zero, the fixed ring is factorial by Proposition 16.4.

Bertin (nee Dumas) [1965] gives an outline for determining the fixed ring of σ. Let B denote the ring $k[X_1, \ldots, X_r]$ and let A denote the fixed ring B^σ. As in Example 16.5, we want to determine B^σ by generators and the ideal of relations. There are three steps one can use after a candidate has been found. First show that the candidate's field of quotients is L^σ. Second show that B is integral over the candidate. Third show that the candidate itself is normal.

The following list of examples is found in [Bertin, 1967]. The notation follows that given above.

Example 16.6. The dimension $r=2$. Then $\sigma = \begin{pmatrix} 1 & 0 \\ 1 & 1 \end{pmatrix}$. So $X_1 \mapsto X_1 + X_2$ and $X_2 \mapsto X_2$. Let u be the product $\prod\limits_{n=1}^{p} (X_1 + n X_2)$.

Then $\sigma(u) = u$. The fixed ring is just $k[X_2, u]$. Since X_2 and u are algebraically independent this ring is clearly normal and factorial.

Example 16.7. Suppose the characteristic $p=2$ and the dimension $r=3$. Then

$$\sigma = \begin{pmatrix} 1 & 0 & 0 \\ 1 & 1 & 0 \\ 0 & 1 & 1 \end{pmatrix}.$$

and $\sigma(X_1) = X_1 + X_2$, $\sigma(X_2) = X_2 + X_3$ and $\sigma(X_3) = X_3$. Let the elements u_1, u_2, u_3 and v_3 be defined by the equations

$$
\begin{aligned}
u_1 &= X_3 \\
u_2 &= X_2(X_2 + X_3) = X_2 \sigma(X_2) \\
u_3 &= X_1(X_1 + X_2)(X_1 + X_3)(X_1 + X_2 + X_3) = X_1 \sigma(X_1) \sigma^2(X_1) \sigma^3 X_1. \\
v_3 &= X_1 X_3(X_1 + X_3) + X_2^2(X_2 + X_3).
\end{aligned}
$$

Now X_3 is fixed by σ. Furthermore $\sigma^3 + \sigma^2 + \sigma + 1 = 0$. Since $X_2 = X_1 + \sigma X_1$, it follows that $\sigma^2 X_2 = X_2$. Hence $X_2 \sigma(X_2)$ is fixed by σ. The element u_3 is the norm of X_1 which is fixed by σ. A quick calculation shows that v_3 is fixed by σ. Moreover

$$v_3^2 + u_1 u_2 v_3 + (u_2^3 + u_3 u_1^2) = 0.$$

The fixed ring is the ring $k[u_1, u_2, u_3, v_3]$, which is factorial.

Example 16.8. Suppose the characteristic $p=2$ and the dimension $r=4$. Then.

$$\sigma = \begin{pmatrix} 1 & 0 & 0 & 0 \\ 1 & 1 & 0 & 0 \\ 0 & 1 & 1 & 0 \\ 0 & 0 & 1 & 1 \end{pmatrix}.$$

We now change the variables. Set $Y_1 = X_1$, $Y_2 = \sigma(X_1)$, $Y_3 = \sigma^2 X_1$ and $Y_4 = \sigma^3 X_1$. Then $Y_2 = X_1 + X_2$, $Y_3 = X_1 + X_3$ and $Y_4 = X_1 + X_2 + X_3 + X_4$.

So $B=k[Y_1,Y_2,Y_3,Y_4]$. Now the matrix of σ is

$$\begin{pmatrix} 0 & 0 & 0 & 1 \\ 1 & 0 & 0 & 0 \\ 0 & 1 & 0 & 0 \\ 0 & 0 & 1 & 0 \end{pmatrix}.$$

Let u_1,u_2,u_3,u_4, x and y be elements in B defined by

$$u_1 = X_4 = Y_1 + \sigma Y_1 + \sigma^2 Y_1 + \sigma^3 Y_1.$$
$$u_2 = X_3(X_3+X_4) = Y_1 Y_2 + Y_2 Y_3 + Y_3 Y_4 + Y_4 Y_1 = (Y_1 + Y_3)(Y_2 + Y_4).$$
$$u_3 = X_2(X_2+X_4)(X_2(X_2+X_4) + X_3(X_3+X_4))$$
$$= (Y_1+Y_2)(Y_3+Y_4)\sigma(Y_1+Y_2)\sigma(Y_3+Y_4).$$
$$u_4 = X_1(X_1+X_2)(X_1+X_3)(X_1+X_2+X_3+X_4) = Y_1\sigma(Y_1)\sigma^2(Y_1)\sigma^3(Y_1).$$
$$x = X_2 X_4(X_2+X_4) + X_3^2(X_3+X_4) = \sum_{i=0}^{3} \sigma^i(Y_1^2(Y_2+Y_3)).$$
$$y = X_1 X_4 + X_2(X_2+X_3+X_4) = Y_1 Y_3 + Y_2 Y_4.$$

It is clear that these elements are fixed by σ. But furthermore

$$x^2 + u_1 u_2 x + u_2^3 + u_1^2 u_3 = 0 \quad \text{and}$$
$$y^4 + u_1^2 y^3 + u_1^2 u_2 y^2 + u_1^2 u_3 y + u_1^4 u_3 + u_3^2 = 0.$$

The ring $k[u_1,u_2,u_3,u_4,x]$ is integrally closed in its field of quotients K'. The invariant subring of B is the integral closure of $k[u_1,u_2,u_3,u_4,x,y]$ in the field $K'[y]$.

The elements u_1,u_2,u_3,u_4 form a system of homogeneous parameters in the graded ring B^σ. For the elements

$$z = u_2 X_1^2 + u_3 X_1 + X_2^2(X_2+X_3+X_4)(X_2+X_4) \quad \text{and}$$
$$t = X_1^2 X_4 + X_2 X_3(X_2+X_3+X_4) = \sum \sigma^i(Y_1 Y_2 Y_3)$$

we have the relations

$$u_1 t + y^2 = u_3$$

and

$$u_1 z + u_2 t + xy = 0.$$

Hence

$$(yz+xt)u_1 + yt u_2 - x u_3 = 0.$$

But neither x nor u_3 lie in the ideal generated by u_1 and u_2. Thus u_1,u_2,u_3,u_4 is not a regular B-sequence. So B^σ cannot be a Cohen-Macaulay ring in the graded sense or, equivalently, that the local ring at the origin is not Cohen-Macaulay.

This example helps to answer a question raised by Peskine and Szpiro [1968]. Suppose C is a local ring of dimension n. If the local cohomology module $H_m^n(C)$ is injective, is the ring C Gorenstein? If R is a regular local ring with C as a factor ring and if $\Omega = \operatorname{Ext}_R^d(C, R)$ where $d = \dim R - \dim C$, then $H_m^n(C) = \operatorname{Hom}_C(\Omega, E)$, the module E being the injective envelope of the residue class field. If C is a normal integral domain, then Ω is isomorphic to a reflexive ideal. In the case that C is factorial, then $H_m^n(C) = E$. But the local ring at the origin in Example 16.8 above is a factor ring of a regular local ring which is not even Cohen-Macaulay let alone Gorenstein.

By using geometrical arguments and the concept of parafactorial rings (see Section 18), it has been shown that geometric factorial rings defined over fields of characteristic zero are Cohen-Macaulay and hence Gorenstein in codimension 3. By refining the argument, one is able to show that a factorial algebra of dimension 4 is Cohen-Macaulay.

§ 17. Radical Descent

In this paragraph we study the fixed ring (i.e., the kernel) of a derivation acting on a Krull domain with characteristic $p \neq 0$. Samuel has used the technique in order to give a large class of examples of complete local Krull domains for which power series extension does not preserve the divisor class group. These examples appear in this section. Danilov [1968, 1970a, 1970b] has done further work on this problem and a discussion of some of his results appears in Section 19. In order that the theory should imitate the theory of Galois descent and in order to be able to handle a finite set of derivations rather than a single derivation, we use the techniques of Zinn-Justin [1965, 1967]. In an appendix we allude to Waterhouse's [1971] work which encompasses both theories into one by considering Hopf algebras and module algebras over them.

Throughout this section, the characteristic of the rings is assumed to be non-zero, as usual denoted by p.

Let L be the field of quotients of the Krull domain B. The ring of dual numbers $L[\delta]$ is the ring $L[t]/(t^2)$. The image of t in $L[\delta]$ is denoted by δ, so that $\delta^2 = 0$. It contains $B + L\delta$ and $B + B\delta$ as subrings. The ring $L[\delta]$ has an augmentation $\pi : L[\delta] \to L$ by $x + y\delta \mapsto x$. The set of ring homomorphisms $\{\alpha : B \to B[\delta] : \pi\alpha = \operatorname{id}\}$ is in bijective correspondence with the module of derivations $\operatorname{Der}(B, B)$.

Let $D_1, \ldots, D_s$ be a finite set of derivations from B to B. Each derivation can be uniquely extended to L. Conversely we can think of each derivation as being given on L but such that it takes B into B. The set $\operatorname{Ker} D_i = \{x \in L : D_i x = 0\}$ is a subfield of L. The intersection

$B \cap \mathrm{Ker}\, D_i = \{b \in B : D_i b = 0\}$ is a Krull domain. Since $D_i x^p = 0$ for all x in L, the field $L^p \subseteq \mathrm{Ker}\, D_i$ and $B^p \subseteq B \cap \mathrm{Ker}\, D_i$. Let K denote the field $\bigcap\limits_{i=1}^{s} \mathrm{Ker}\, D_i$ and denote by A the Krull domain $B \cap K$. Then B is integral over A so there are induced homomorphisms $\mathrm{Div}(A) \to \mathrm{Div}(B)$ and $i : \mathrm{Cl}(A) \to \mathrm{Cl}(B)$. Once again we wish to compute the kernel of $i : \mathrm{Cl}(A) \to \mathrm{Cl}(B)$.

Rather than considering the set $\{\alpha : L \to L[\delta] : \pi\alpha = \mathrm{id}_L\}$ we consider each derivation as an automorphism of $L[\delta]$. For a derivation d of L, the assignment $x + y\delta \mapsto x + (d(x) + y)\delta$ for all x and y in L defines an automorphism of $L[\delta]$. We denote this automorphism by $\alpha(d)$. If d' is another derivation, then $\alpha(d + d') = \alpha(d)\alpha(d')$ and $\alpha(d)^{-1} = \alpha(-d)$. Thus we obtain a homomorphism $\mathrm{Der}(L, L) \to \mathrm{Aut}(L[\delta])$ which is a monomorphism. But notice that $\pi\alpha(d) = \pi$. Let $\mathrm{Aut}_\pi(L[\delta])$ denote the subgroup of $\mathrm{Aut}(L[\delta])$ of those automorphism α such that $\pi\alpha = \pi$. There is a surjection $\mathrm{Aut}_\pi(L[\delta]) \to L^*\delta$ given by $\alpha \mapsto \alpha(\delta)$ whose kernel is exactly $\mathrm{Der}(L, L)$. So the sequence

$$0 \to \mathrm{Der}(L, L) \to \mathrm{Aut}_\pi(L[\delta]) \to L^*\delta \to 1$$

is exact. It is even split in the sense of group extension. For it is easy to see that such an automorphism α is of the form $\alpha(x + y\delta) = x + d(x)\delta + y\alpha(\delta)$. Let $u\delta = \alpha(\delta)$ and denote this automorphism by (d, u). Then the composition $(d, u) \cdot (d', v)$ is the automorphism $(d + ud', uv)$. This is the product in the semi-direct product.

The group $\mathrm{Der}(L, L)$ is p-torsion, so any finite set of elements generates a finite subgroup of $\mathrm{Der}(L, L)$ or equivalently a finite subgroup of $\mathrm{Aut}_\pi(L[\delta])$ which is a p-group. If a derivation preserves B, then it induces, in the same way, a π-automorphism of the ring $B + L\delta$ (and of $B[\delta]$).

Let G be a finite group of derivations of B which we consider also as automorphisms of $L[\delta]$ and $B + L\delta$. Then each element of G induces an automorphism of the respective groups of units $L[\delta]^*$ and $(B + L\delta)^*$. But these groups are just $L^* + L\delta$ and $B^* + L\delta$ respectively. Therefore we get the exact sequence

$$1 \to B^* + L\delta \to L^* + L\delta \to \mathrm{Prin}\, B \to 0,$$

each term being a G-module, the action on $\mathrm{Prin}\, B$ being trivial.

Let A be the kernel of the elements in G with field of quotients K. Then the invariant subgroups are $A^* + L\delta$ and $K^* + L\delta$ respectively. Therefore the long exact sequence of cohomology is

$$1 \to A^* + L\delta \to K^* + L\delta \to \mathrm{Prin}\, B \to \mathrm{H}^1(G, B^* + L\delta) \to \cdots .$$

The cokernel of $A^* + L\delta \to K^* + L\delta$ is $\operatorname{Prin} A$, so the exact sequence then fits into a large diagram with exact rows and columns.

$$
\begin{array}{ccccccc}
& & 0 & & 0 & & \\
& & \downarrow & & \downarrow & & \\
0 & \to & \operatorname{Prin} A & \to & \operatorname{Prin} B & \to & H^1(G, B^* + L\delta) & \to & H^1(G, L^* + L\delta) & \to \\
& & \downarrow & & \downarrow & & \\
& & \operatorname{Div} A & \to & \operatorname{Div} B & & & \to & H^1(G, \operatorname{Prin} B) \\
& & \downarrow & & \downarrow & & \\
& & \operatorname{Cl}(A) & \xrightarrow{i} & \operatorname{Cl}(B) & & \\
& & \downarrow & & \downarrow & & \\
& & 0 & & 0 & & .
\end{array}
$$

The homomorphism $\operatorname{Div} A \to \operatorname{Div} B$ is an injection. Thus there is induced a homomorphism $\theta_G : \operatorname{Ker} i \to H^1(G, B^* + L\delta)$ which is an injection. In order to continue further we want to determine the groups $H^1(G, B^* + L\delta)$ and $H^1(G, L^* + L\delta)$. But let us collect the present material.

Proposition 17.1. *Let G be a finite group of derivations of the Krull domain B. Let A be the fixed subring of G. Then there is an exact sequence of groups*

$$0 \to \operatorname{Prin} A \to \operatorname{Prin} B \to H^1(G, B^* + L\delta) \to H^1(G, L^* + L\delta) \to 0$$

which induces an injection

$$\theta : \operatorname{Ker} i \to H^1(G, B^* + L\delta).$$

Proof. The only thing left to show is that $H^1(G, B^* + L\delta) \to H^1(G, L^* + L\delta)$ is onto. But $H^1(G, \operatorname{Prin} B) = \operatorname{Hom}_{\mathbb{Z}}(G, \operatorname{Prin} B)$ since G acts trivally on $\operatorname{Prin} B$. But G is a torsion group while $\operatorname{Prin} B$ is torsion free, it being a subgroup of the free group $\operatorname{Div} B$. Thus $H^1(G, \operatorname{Prin} B) = 0$. □

Let r be the rank of the group G as a vector space over the field $\mathbb{Z}/p\,\mathbb{Z}$. So G is generated by r independent derivations $D_1, \ldots, D_r$ with corresponding automorphisms $\alpha_1, \ldots, \alpha_r$. Let V be the vector space L^r. Denote by V_0 the subgroup of V defined by $V_0 = \{(u^{-1}D_1 u, \ldots, u^{-1}D_r u) : u \in B^*\}$. Let V_1 be the subgroup $\{(x^{-1}D_1 x, \ldots, x^{-1}D_r x) : x \in L^*\}$.

Lemma 17.2. *There are isomorphisms $H^1(G, B^* + L\delta) \cong V/V_0$ and $V/V_1 \cong H^1(G, L^* + L\delta)$ so that the natural homomorphism $H^1(G, B^* + L\delta) \to H^1(G, L^* + L\delta)$ is the induced homomorphism $V/V_0 \to V/V_1$.*

Proof. The cohomology group $H^1(G, L^* + L\delta)$ is the group of 1-cocycles modulo 1-coboundaries. A 1-cocycle $f : G \to B^* + L\delta$ is determined by the images of the automorphisms $f(\alpha_i)$; $i = 1, 2, \ldots, r$ and by the relation $f(\alpha \beta) = \alpha f(\beta) f(\alpha)$ for α and β in G.

Let α be an element in G. Let $f(\alpha)=(u+x\delta)$ where $u\in L^*$ and $x\in L$. Then $\alpha f(\alpha)=\alpha(u+x\delta)=u+(d(u)+x)\delta$, the d being the derivation associated to α. By induction, the powers of α on $f(\alpha)$ are given by $\alpha^r f(\alpha)=f(\alpha)+rd(u)\delta$. By induction on the cocycle relation we see that $f(\alpha^r)=\prod_{j=0}^{r}\alpha^j f(\alpha)$. Hence $f(\alpha^r)=\prod_{j=0}^{r}(f(\alpha)+jd(u)\delta)$ which is $f(\alpha)^r$ $+\binom{r+1}{2}f(\alpha)^{r-1}d(u)\delta$. Now $f(\alpha)=u+x\delta$ and so $f(\alpha)^r=u^r+ru^{r-1}x\delta$ while $f(\alpha)^{r-1}d(u)\delta=u^{r-1}d(u)\delta$. Hence $f(\alpha^r)=u^r+u^{r-1}\left(rx+\binom{r+1}{2}d(u)\right)\delta$.

Since $\alpha^{p+1}=\alpha$, this relation yields $u+x\delta=u^{p+1}+u^p(x+d(u))\delta$. But then $u=1$. Hence $f(\alpha)=1+x\delta$ for any 1-cocycle f. Furthermore $f(\alpha\beta)=f(\alpha)f(\beta)$. Hence a 1-cocycle is a homomorphism. Therefore the group of 1-cocycles is just the group L' (we can define f by $f(\alpha_1)=1+x_1\delta$ and so on).

A 1-coboundary is a 1-cocycle f of the form $f(\alpha)=(u+x\delta)^{-1}\alpha(u+x\delta)$ for some $u+x\delta$ in $L^*+L\delta$ and for each α in G. Now $(u+x\delta)^{-1}$ $=u^{-1}-u^{-2}x\delta$. Hence $f(\alpha)=(u^{-1}-u^{-2}x\delta)(u+(d(u)+x)\delta)$. Thus $f(\alpha)=1+u^{-1}d(u)\delta$. The element in L' corresponding to this 1-coboundary is the vector $(u^{-1}D_1(u),\ldots,u^{-1}D_r(u))$ for u in L^*. (Note that $u^{-1}D(u)+v^{-1}D(v)=(uv)^{-1}D(uv)$.) Thus $H^1(G,L^*+L\delta)=V/V_1$.

The same calculations yield the isomorphism $H^1(G,B^*+L\delta)=V/V_0$. The last statement is clear. $\quad\square$

Corollary 17.3. *The homomorphism defined by* $xB\mapsto(x^{-1}D_1(x),\ldots,$ $x^{-1}D_r(x))$ *induces an exact sequence*

$$0\to \operatorname{Prin}A\to \operatorname{Prin}B\to V_1/V_0\to 0.$$

The kernel of the homomorphism $\operatorname{Cl}(A)\to\operatorname{Cl}(B)$ *is isomorphic to a subgroup of* V_1/V_0. *In fact, the kernel is isomorphic to a subgroup of* V_0'/V_0 *with* $V_0'=\{(u^{-1}D_1u,\ldots,u^{-1}D_ru):u^{-1}D_iu\in B$ *for all* $i\}$.

Proof. We have verified the first two statements. As for the third, suppose $\mathfrak{a}$ is a divisorial ideal of A whose class is in the kernel. Then $B:(B:\mathfrak{a}B)$ is a principal ideal, say $B:(B:\mathfrak{a}B)=xB$ for some x in B. For each $\mathfrak{P}$ in $X^{(1)}(B)$, the ideal $xB_{\mathfrak{P}}=\mathfrak{a}B_{\mathfrak{P}}$. Thus there is an element $a(\mathfrak{P})$ in $\mathfrak{a}$ such that $xB_{\mathfrak{P}}=a(\mathfrak{P})B_{\mathfrak{P}}$. Let u be a unit in $B_{\mathfrak{P}}$ such that $x=u\cdot a(\mathfrak{P})$. Then $D_i(x)=a(\mathfrak{P})D_i(u)$ for each $i=1,2,\ldots,r$. But then $x^{-1}D_i(x)=u^{-1}D_i(u)$ for each i. Thus $x^{-1}D_i(x)\in B_{\mathfrak{P}}$ for all $\mathfrak{P}$ in $X^{(1)}(B)$. Hence the element $(x^{-1}D_1(x),\ldots,x^{-1}D_r(x))\in V_0'$. But this is just the image of $\operatorname{cl}(\mathfrak{a})$ in V_1/V_0. $\quad\square$

Yuan [1968] states conditions for the injection $\operatorname{Ker} i \to V_0'/V_0$ to be a bijection. The derivations of the group G are endomorphisms of B. Let $B[G]$ denote the A-subalgebra of $\operatorname{End}_A(B)$ generated by B and G. Yuan's result then is the following theorem.

Theorem 17.4. *Suppose L is a finite extension of K and that B is an A-lattice in L. If $B[G] = \operatorname{End}_A(B)$, then $\operatorname{Ker} i \to V_0'/V_0$ is a bijection.*

Proof. It remains to show that the map is a surjection. Suppose x is an element whose derived logarithm $x^{-1} D_i x \in B$ for each i. Let $\operatorname{div}(x B)$ be the sum $\sum_{\mathfrak{P}} v_{\mathfrak{P}}(x) \operatorname{div} \mathfrak{P}$. Now $x B$ is the image of an extended ideal $\mathfrak{a}$ if each $v_{\mathfrak{P}}(x)$ is divisible by the ramification index of $\mathfrak{P}$ over A. Since $B^p \subseteq A$, the only possibilities for the ramification index are p and 1.

Suppose t is a uniformizing parameter for $B_{\mathfrak{P}}$. Then there is a unit u in $B_{\mathfrak{P}}$ such that $x = u t^{v(x)}$, the valuation v being $v_{\mathfrak{P}}$. Hence $x^{-1} D_i x = u^{-1} D_i u + v(x) t^{-1} D_i t$. If $v_{\mathfrak{P}}(x) \cdot 1$ is a unit in $B_{\mathfrak{P}}$, then the element $t^{-1} D_i t$ must lie in $B_{\mathfrak{P}}$. Hence $D_i(\mathfrak{P} B_{\mathfrak{P}}) \subseteq \mathfrak{P} B_{\mathfrak{P}}$ for each i.

Continue the assumption that $v_{\mathfrak{P}}(x) \cdot 1$ is a unit in $B_{\mathfrak{P}}$. Let $\mathfrak{p}$ denote the prime ideal $\mathfrak{P} \cap A$. Since $B^p \subseteq A$, the rings of quotients $A_{\mathfrak{p}} \otimes_A B$ and $B_{\mathfrak{P}}$ coincide. Hence $A_{\mathfrak{p}} \otimes_A B[G] = B_{\mathfrak{P}}[G]$. Since B is an A-lattice it is a divisorial A-lattice. Thus $A_{\mathfrak{p}} \otimes_A \operatorname{Hom}_A(B, B) = \operatorname{Hom}_{A_{\mathfrak{p}}}(B_{\mathfrak{P}}, B_{\mathfrak{P}})$ for each $\mathfrak{P}$ in $X^{(1)}(B)$. By hypothesis $B[G] = \operatorname{Hom}_A(B, B)$ and hence $B_{\mathfrak{P}}[G] = \operatorname{Hom}_{A_{\mathfrak{p}}}(B_{\mathfrak{P}}, B_{\mathfrak{P}})$ for each $\mathfrak{P}$ in $X^{(1)}(B)$.

Let $\mathscr{L}(G)$ denote the restricted Lie algebra of K-derivations of L. Then $[\mathscr{L}(G):L] = m$ where $[L:K] = p^m$ [Jacobson, 1964]. Since $L[G] = \operatorname{Hom}_K(L, L)$, the algebra $\mathscr{L}(G)$ is generated by the derivations in G.

Since $B_{\mathfrak{P}}[G] = \operatorname{Hom}_{A_{\mathfrak{p}}}(B_{\mathfrak{P}}, B_{\mathfrak{P}})$, the restricted Lie $B_{\mathfrak{P}}$-algebra $\mathscr{L}_{B_{\mathfrak{P}}}[G]$ of all $A_{\mathfrak{p}}$-derivations of $B_{\mathfrak{P}}$ has rank m as a left $B_{\mathfrak{P}}$-module. Each element in $B_{\mathfrak{P}}[G]$ leaves fixed the maximal ideal so there is induced a homomorphism $\mathscr{L}_{B_{\mathfrak{P}}}[G] \to \mathscr{L}_{k(\mathfrak{P})}[\bar{G}]$ which factors through $k(\mathfrak{P}) \otimes_{B_{\mathfrak{P}}} \mathscr{L}_{B_{\mathfrak{P}}}[G]$.

Suppose D is an $A_{\mathfrak{p}}$-derivation of $B_{\mathfrak{P}}$ which has the property that $D(B_{\mathfrak{P}}) \subseteq \mathfrak{P} B_{\mathfrak{P}}$. Then $t^{-1} D$ is a derivation of L which preserves $B_{\mathfrak{P}}$. Hence there are elements $b_1, \ldots, b_r$ in $B_{\mathfrak{P}}$ such that $t^{-1} D = b_1 D_1 + \cdots + b_r D_r$ and therefore $D = t b_1 D_1 + \cdots + t b_r D_r$, which is an element in $\mathfrak{P} \mathscr{L}_{B_{\mathfrak{P}}}[G]$. So we have seen that any derivation which induces the zero derivation on $k(\mathfrak{P})$ lies in $\mathfrak{P} \mathscr{L}_{B_{\mathfrak{P}}}[G]$. Therefore $k(\mathfrak{P}) \otimes \mathscr{L}_{B_{\mathfrak{P}}}[G] \to \mathscr{L}_{k(\mathfrak{P})}[\bar{G}]$ is a monomorphism. Thus the dimension $[\mathscr{L}_{k(\mathfrak{P})}[\bar{G}] : k(\mathfrak{P})] \geqslant [\mathscr{L}[G] : L]$.

Thus if every derivation in G preserves the maximal ideal of $B_{\mathfrak{P}}$, then $[k(\mathfrak{P}):k(\mathfrak{p})] \geqslant [L:K]$. The reverse inequality also holds, since $[k(\mathfrak{P}):k(\mathfrak{p})] \leqslant [B_{\mathfrak{P}}/\mathfrak{p} B_{\mathfrak{P}} : k(\mathfrak{p})]$. Hence $[k(\mathfrak{P}):k(\mathfrak{p})] = [L:K]$. Then from the well known inequality $e_{\mathfrak{P}}[k(\mathfrak{P}):k(\mathfrak{p})] \leqslant [L:K]$ we conclude $e_{\mathfrak{P}} = 1$, where $e_{\mathfrak{P}}$ is the index of ramification of $\mathfrak{P}$.

Thus the principal fractionary ideal of B generated by an element whose derived logarithms lie in B is the extension of an ideal of A. $\square$

Remark 1. The proof shows that a prime ideal $\mathfrak{P}$ in $X^{(1)}(B)$ with the property that $D_i(\mathfrak{P}) \subseteq \mathfrak{P} B_{\mathfrak{P}}$ for each i cannot be ramified over A. The converse is also true. For suppose $\mathfrak{P}$ is unramified over A. Then $\mathfrak{P} B_{\mathfrak{P}} = t B_{\mathfrak{P}}$ for some t in A. Then $D_i(\mathfrak{P} B_{\mathfrak{P}}) \subseteq \mathfrak{P} B_{\mathfrak{P}}$ for each i.

Remark 2. The condition that B be a divisorial A-lattice is only to insure that $\operatorname{Hom}_A(B,B)_{\mathfrak{p}} = \operatorname{Hom}_{A_{\mathfrak{p}}}(B_{\mathfrak{p}}, B_{\mathfrak{p}})$ for each $\mathfrak{p}$ in $X^{(1)}(A)$. It is enough to assume that $B_{\mathfrak{P}}[G] = \operatorname{Hom}_{A_{\mathfrak{p}}}(B_{\mathfrak{P}}, B_{\mathfrak{P}})$ for each $\mathfrak{P}$ in $X^{(1)}(B)$. Waterhouse [1971] also insists on such a condition. His result follows in case B is an A-lattice, since the localization at $\mathfrak{p}$ in $X^{(1)}(A)$ commutes with Hom_A as is seen in Corollary 5.5.

The point to Remark 2 is that Samuel [1964 b] has demonstrated a slightly different result.

Proposition 17.5. *Suppose* $[L:K] = p$ *and that* $D(B)$ *is contained in no prime divisorial ideal of* B. *Then* $\operatorname{Ker} i \to V'_0/V_0$ *is a bijection.*

Proof. The proof is the same as the proof of 17.4 up to the point where the derivation $\bar{D}: k(\mathfrak{P}) \to k(\mathfrak{P})$ is defined. Since there is an element b in B with Db not in $\mathfrak{P}$, the derivation $\bar{D} \neq 0$. Hence $[k(\mathfrak{P}):k(\mathfrak{p})] = p$ and then the ramification index must be one. $\square$

There is no assumption that B be an A-lattice in L. However $B_{\mathfrak{P}}[D] = \operatorname{Hom}_{A_{\mathfrak{p}}}(B_{\mathfrak{P}}, B_{\mathfrak{P}})$ for every prime $\mathfrak{P}$, for, as we will show, the minimal polynomial for D over K is $t^p - at$ for some a in A. Thus $B_{\mathfrak{P}}[D]$ is free of rank p over $B_{\mathfrak{P}}$ (as a left $B_{\mathfrak{P}}$-module). Hence $B_{\mathfrak{P}}[D] = \operatorname{Hom}_{A_{\mathfrak{p}}}(B_{\mathfrak{P}}, B_{\mathfrak{P}})$ by a rank argument.

The assumption $B[G] = \operatorname{Hom}_A(B,B)$ cannot be relaxed too much as is seen by an example due to Samuel [1964 b]. Suppose $\operatorname{char} k = p$ and that $B = k[X,Y,Z]$. Define $D: B \to B$ by $D(X) = Y^4$, $D(Y) = X^2$ and $D(Z) = XYZ$ and extend to B. Then $A = k[X^2, Y^2, Z^2]$ is a factorial ring. So $\ker i = 0$ while $V'_0/V_0 \neq (0)$ (since $Z^{-1} DZ \in B$ but $Z \notin V_0$).

The next result will be very useful in discussing the examples.

Lemma 17.6. *Let* $D: L \to L$ *be a derivation with kernel* K. *Suppose* $[L:K] = p$.
(a) *There is an element* a *in* A *such that* $D^p = aD$.
(b) *An element* z *in* L *is a logarithmic derivative if and only if*

$$D^{p-1} z - az + z^p = 0.$$

Proof. The statements in this lemma are special cases of some exercises in [Jacobson, 1964]. There the assumption is that $[L:K] = p^m$

where $K = \operatorname{Ker} D$. Then there is a p-polynomial $f(t) = t^{p^m} + b_1 t^{p^{m-1}} + \cdots + b_m t$ in $K[t]$ such that $f(D) = 0$.

For each x in L, define $x^{[p^j]}$ to be the element

$$x^{[p^j]} = \sum_{i=0}^{j} (D^{p^i-1} x)^{p^{j-i}}.$$

Then x is a logarithmic derivative if and only if

$$x^{[p^m]} + b_1 x^{[p^{m-1}]} + \cdots + b_{m-1} x^{[p]} + b_m x = 0.$$

This is the differential analogue of Hilbert's Theorem 90. There is also an analogue used in Waterhouse's paper.

There is only one thing left to prove in the statement of the lemma; that the coefficient in the polynomial for D is in A.

By the Jacobson exercise, there is an a in K such that $D^p = aD$. Suppose $a \notin A$. Then there is a prime divisorial ideal $\mathfrak{P}$ in $X^{(1)}(B)$ such that $v_{\mathfrak{P}}(a) < 0$. Suppose t is a uniformizing parameter in $B_{\mathfrak{P}}$. Then $D^p(t) = aD(t)$. Since $D(t) \in B_{\mathfrak{P}}$, the value $v_{\mathfrak{P}}(D^p(t)) \geqslant 0$. But $v_{\mathfrak{P}}(D^p(t)) = v_{\mathfrak{P}}(a) + v_{\mathfrak{P}}(D(t))$. Hence $v_{\mathfrak{P}}(D(t)) > 0$. For any x in L, there is a unit u in $B_{\mathfrak{P}}$ and an integer r such that $x = t^r u$. Then $Dx = r t^{r-1} D(t) u + t^r Du$. We factor t^r from this term on the right to get $Dx = t^r (r u t^{-1} D(t) + D(u))$. Since $D(t) \in t B_{\mathfrak{P}}$, this element is in $t^r B_{\mathfrak{P}}$. Hence $v_{\mathfrak{P}}(D(x)) \geqslant v_{\mathfrak{P}}(x)$ for all x in L. By induction $v_{\mathfrak{P}}(D^n(x)) \geqslant v_{\mathfrak{P}}(x)$. In particular, the value $v_{\mathfrak{P}}(D^{p-1}(D(t))) \geqslant v_{\mathfrak{P}}(D(t))$. Hence $v_{\mathfrak{P}}(a) + v_{\mathfrak{P}}(D(t)) \geqslant v_{\mathfrak{P}}(D(t))$ which shows that $v_{\mathfrak{P}}(a) \geqslant 0$, a contradiction. Hence $a \in B$ and $a \in K$, so $a \in A$. ☐

Now these results are applied to obtain some more examples of non-factorial rings. The game is the same as for Galois descent. One is given a factorial ring on which a derivation can be defined. The fixed ring is determined and the group V_0'/V_0 is determined.

Let $\mathfrak{o}$ be a Krull domain with field of quotients k. Let B be the Krull domain $\mathfrak{o}[X, Y]$ with field of quotients L. Let further $D: L \to L$ be a k-derivation such that $D(B) \subseteq B$. Since the group of units of B is $\mathfrak{o}^*$, the group of units of $\mathfrak{o}$, the group of derived logarithms of units of B is just (0). If A denotes $B \cap \operatorname{Ker} D$, then $\operatorname{Cl}(A)$ is isomorphic to a subgroup of V_0'. In case the assumptions in Theorem 17.4 or Proposition 17.5 are fulfilled, then $\operatorname{Cl}(A) \cong V_0'$.

Example 17.7. Define $D: L \to L$ by setting $DX = X$, $DY = -Y$ and extending to all of L. Then $k[X^p, Y^p, XY] \subseteq \operatorname{Ker} D$, so the field of quotients of $k[X^p, Y^p, XY] \subseteq \operatorname{Ker} D$. Now $[L : k(X^p, Y^p, XY)] = p$, so the field of quotients of $k[X^p, Y^p, XY] = \operatorname{Ker} D$. It is easy to see that $k[X^p, Y^p, XY] = k[U, V, W]/(W^p - UV)$, and that the surface $W^p = UV$ is non-singular away from the origin. Hence $k[X^p, Y^p, XY]$ is normal. Now $\mathfrak{o}[X^p, Y^p, XY] = k[X^p, Y^p, XY] \cap \mathfrak{o}[X, Y]$. Hence $\mathfrak{o}[X^p, Y^p, XY]$ is just $(\operatorname{Ker} D) \cap B$.

The extensions $\mathfrak{o} \to B$ and $A \to B$ satisfy the condition (PDE). Since B is integral over A, the extension $\mathfrak{o} \to A$ satisfies (PDE). Thus we get the homomorphisms

$$\mathrm{Cl}(\mathfrak{o}) \to \mathrm{Cl}(A) \to \mathrm{Cl}(B)$$

whose composition is the bijection

$$\mathrm{Cl}(\mathfrak{o}) \to \mathrm{Cl}(B).$$

Thus $\mathrm{Cl}(A) \to \mathrm{Cl}(B)$ is onto and splits.

Since $DB \supseteq BX + BY$ and $[L:K] = p$, the hypotheses of Proposition 17.5 are fulfilled. So $\mathrm{Ker}(\mathrm{Cl}(A) \to \mathrm{Cl}(B)) = V_0'$ and we get the isomorphism

$$\mathrm{Cl}(\mathfrak{o}[X^p, Y^p, XY]) = \mathrm{Cl}(\mathfrak{o}) \oplus V_0'.$$

It remains to determine V_0'. Let x in L^* be an element whose derived logarithm $x^{-1} Dx$ is in B. We can write $x = b/a$ where $b \in B$ and $a \in A$. Then $x^{-1} Dx = b^{-1} Db$, For each polynomial b in B, the degree $\deg Db \leqslant \deg b$. If $b^{-1} Db \in B$, then $Db = bu$ where $u \in \mathfrak{o}$. But then $D^n b = bu^n$ for all $n \geqslant 0$. Since $D^p = D$, we have $bu^p = bu$. As $b \neq 0$, we get $u^p = u$. But this implies that u is one of the p elements in the rational field $\mathbb{F}_p \subseteq \mathfrak{o}$. On the other hand, for each $n \geqslant 0$, we have $D(X^{n+1} Y) = n X^{n+1} Y$. So if u is the image in $\mathbb{F}_p$ of n in $\mathbb{Z}$, $n \geqslant 0$, then $D X^{n+1} Y = u X^{n+1} Y$. Thus we get $\mathbb{F}_p \cong V_0'$. Substituting this into the isomorphism above we get $\mathrm{Cl}(\mathfrak{o}[X^p, Y^p, XY]) = \mathrm{Cl}(\mathfrak{o}) \oplus \mathbb{F}_p$. For instance, when $\mathfrak{o}$ is factorial, then the class group $\mathrm{Cl}(\mathfrak{o}[X^p, Y^p, XY])$ is cyclic of order p. This is the same result as that obtained for Galois descent.

Example 17.8. Suppose i and j are integers, both relatively prime to p. Define $D: L \to L$ by extending the function $DX = j Y^{j-1}$ and $DY = -i X^{i-1}$ to all of L. Just as in the previous case, one shows that $\mathrm{Ker} D = k(X^p, Y^p, X^i + Y^j)$ and that $A = \mathfrak{o}[X^p, Y^p, X^i + Y^j] \cong \mathfrak{o}[U, V, W]/(W^p - U^i - V^j)$. Also $\mathrm{Cl}(A) \to \mathrm{Cl}(B) = \mathrm{Cl}(\mathfrak{o})$ is a split surjection. Thus we get the isomorphism $\mathrm{Cl}(A) = \mathrm{Cl}(\mathfrak{o}) \oplus V_0'$. We must determine V_0'.

In fact V_0' is rather difficult to calculate. Attach weights j to X and i to Y respectively. A monomial $X^a Y^b$ then has weight $aj + bi$. In this manner, the ring B becomes a graded ring. Since $D(X^a Y^b) = -bi X^{a+i-1} Y^{b-1} + aj X^{a-1} Y^{b+j-1}$, the derivation D is a homogeneous function and $D(B_n) \subseteq B_{n+ij-i-j}$.

Let a in A be the element such that $D^p = aD$. Then a must be a homogeneous element of degree $(p-1)(ij - i - j)$.

Suppose x in L^* has its derived logarithm $x^{-1} Dx$ in B. Then $x^{-1} Dx = b^{-1} Db$ for some b in B. Let $b^{-1} Db = F$. Write $b = b_d + b_{d+1} + \cdots + b_{d+r}$ and $F = F_h + \cdots + F_{h+s}$ where the b_j and F_i are homogeneous elements, where b_d, b_{d+r}, F_h and F_{h+s} are all different from zero. Then we get $D b_d + \cdots + D b_{d+r} = (b_d + \cdots + b_{d+r})(F_h + \cdots + F_{h+s})$. The right hand side has $b_d F_h$ as its term of lowest weight and $b_{d+r} F_{h+s}$ as its term of highest weight. The weight of the first nonzero term on the left hand side is at least $d + (ij - i - j)$, while the weight of the last nonzero term is at most $d + r + ij - i - j$. Hence we get two inequalities

$$d + ij - i - j \leqslant d + h$$

and

$$d + r + ij - i - j \geqslant d + r + h + s.$$

After simplicifation, these become

$$ij - i - j \leqslant h$$

and $ij - i - j \geqslant h + s$. Since $s \geqslant 0$, these two inequalities imply both $s = 0$ and $h = ij - i - j$. Hence F must be homogeneous of weight $ij - i - j$. Let $X^a Y^b$ be a monomial occurring in F. Then

$$aj + bi = ij - i - j,$$

or

$$(a + 1)j = (j - 1 - b)i.$$

Let d be the greatest common divisor of i and j. Set $i = dr$ and $j = ds$. Then we can cancel d in the equation to get $(a + 1)s = (j - 1 - b)r$. Since s and r are relatively prime, there is an integer n such that

$$a + 1 = nr$$

and

$$j - 1 - b = ns.$$

Hence $a = nr - 1$ and $b = j - ns - 1$. Since both a and b are non-negative, we must have $1 \leqslant n \leqslant d - 1$. On the other hand, any such n gives the two integers a and b and the monomial $X^a Y^b$ with weight $ij - i - j$. So we can write

$$F = \sum_{n=1}^{d-1} a_n X^{nr-1} Y^{j-ns-1},$$

where $a_n \in A$.

Now the element F, being a derived logarithm, satisfies the equation $F^p = a F - D^{p-1} F$ according to Lemma 17.6. Both sides of the equation are forms of weight $p(ij - i - j)$. If one collects the coefficients of a given monomial $X^\alpha Y^\beta$ of the form on the right hand side, one gets a linear form in the coefficients $a_1, \ldots, a_{d-1}$, say $L_{\alpha, \beta}(\mathbf{a})$. On the left hand side one has the form

$$\sum_{n=1}^{d-1} a_n^p X^{p(nr-1)} Y^{p(j-ns-1)}.$$

Thus we get the equations

and
$$a_n^p = L_{p(nr-1),\, p(j-ns-1)}(\mathbf{a})\ \bigl(= L_n(\mathbf{a})\bigr),$$

$$0 = L_{\alpha,\beta}(\mathbf{a})$$

for $n=1,\dots,d-1$, and $\alpha j + \beta i = p(ij-i-j)$, where $\alpha \neq p(nr-1)$ (and so $\beta \neq p(j-ns-1)$).

If $a_1,\dots,a_{d-1}$ are elements of $\mathfrak{o}$ which satisfy these equations, then the corrosponding F is a derived logarithm. (One should note that a is a form of weight $(p-1)(ij-i-j)$ with its coefficients in the ground field $\mathbb{F}_p$, since the derivation is defined over $\mathbb{F}_p$. Thus it does indeed make sense to talk about the forms $L_{\alpha,\beta}(\mathbf{a})$ which are defined over $\mathbb{F}_p$.)

Let $X_0, X_1,\dots, X_{d-1}$ be homogeneous coordinates for $\mathbf{P}^{d-1}(k)$. Then the hypersurfaces

$$X_n^p = L_n(\mathbf{X}) X_0^{p-1} \qquad \bigl(\mathbf{X} = (X_1,\dots, X_{d-1})\bigr),$$

have no solutions at $X_0 = 0$. Hence the set of solutions is a finite set and, by Bezout's theorem, there are at most p^{d-1} solutions in an algebraic closure of k.

Thus V_0' is a finite additive subgroup of B. It must be an elementary p-group of order at most p^{d-1}. If k is algebraically closed and $\mathfrak{o}=k$, then the order is exactly p^{d-1}. If $\mathfrak{o}\neq k$, there can conceivably be solutions in k but not over $\mathfrak{o}$.

Let $\mathrm{P}(i,j)$ denote this group. Then $\mathrm{P}(i,j)$ is an elementary p-group of order at most p^{d-1} with d the greatest common divisor of i and j. The order of $\mathrm{P}(i,j)$ is p^{d-1} provided $\mathfrak{o}$ contains the algebraic closure of $\mathbb{F}_p$.

Let $\mathfrak{o}$ be a Krull domain. Suppose i and j are two integers relatively prime to p. Suppose $A = \mathfrak{o}[U, V, W]/(W^p - U^i - V^j)$. Then $\mathrm{Cl}(A) = \mathrm{Cl}(\mathfrak{o}) \oplus \mathrm{P}(i,j)$.

The group $\mathrm{P}(i,j)$ has been determined in some cases in which the characteristic is 2 or 3. Let us examine more closely the case $p=2$. Since i and j are relatively prime to 2, they and their greatest common divisor d are odd. Write $d=2c+1$. Now $D^2 X = D(j\, Y^{j-1})$ which is $j(j-1)D(Y)$. Since both i and j are odd, the second derivative D^2 is identically zero. So if $D^2 = aD$, then $a=0$. Hence an element F in B is a logarithmic derivative if and only if $F^2 = DF$.

Write $F = \displaystyle\sum_{n=1}^{d-1} a_{n-1} X^{nr-1} Y^{j-ns-1}$ as before (where we have shifted

the vector $\mathbf{a}$ to $(a_0,\dots,a_{d-2})$). Now $D(X^{nr-1} Y^{j-ns-1}) = j(nr-1)X^{nr-2} Y^{2j-ns-2} - i(j-ns-1)X^{nr+i-2} Y^{j-ns-2}$. If n is even this is just $X^{nr-2} Y^{aj-ns-2}$, while if n is odd it is $X^{nr+i-2} Y^{j-ns-2}$ (since i,j,r,s and d are all odd).

Let n be odd, say $n = 2k+1$. Then the term in F^2 corresponding to term $X^{nr+i-2} Y^{j-ns-2}$ is $X^{2(mr-1)} Y^{2(j-ms-2)}$. Hence $2m = n+d$ which is $2k+1+2c+1$. Hence $m = k+c+1$. When the coefficients are attached to the monomials we get $a_{k+c}^2 = a_{2k}$ for $0 \leqslant k \leqslant c-1$. If n is even, then the corresponding equations for the coefficients are $a_{k-1}^2 = a_{2k-1}$ for $1 \leqslant k \leqslant c$. Let S be the set $\mathbb{Z}/d\mathbb{Z} - \{-1\}$. Then $h: S \to S$, defined by $h(s) = 2s+1$, is a permutation of S. The equations then have the form $a_m^2 = a_{h(m)}$ for m in S. Let $\mathcal{O}_1, \ldots, \mathcal{O}_s$ be the orbits of the permutation h with orders $o_1, \ldots, o_s$ respectively. Then $o_1 + \cdots + o_s = \mathrm{card}\, S = d-1 = 2c$. Let $\mathcal{O}_x$ be an orbit, so $\mathcal{O}_x = \{x, h(x), \ldots, h^{r-1}(x)\}$, with r the order. Then $a_{h^j(x)}^2 = a_{h^{j+1}(x)}$ for all j if and only if $a_x^{2^r} = a_x$ and $a_{h(x)} = a_x^2$, etc. The solutions to $a^{2^r} = a$ in $\mathfrak{o}$ are in the subfield $\mathbb{F}_{2^r} \cap \mathfrak{o}$. This is the case for each orbit. Therefore the group of derived logarithms in B is just the direct sum

$$\coprod_{i=1}^{s} (\mathbb{F}_{2^{o_i}} \cap \mathfrak{o}).$$

Note that $\mathbb{F}_{2^{o_i}} \cap \mathfrak{o} \supseteq \mathbb{F}_2$, so that the order of the group $P(i,j)$ is bounded from below by 2^s.

It is possible to make some estimates about the size of the orbits and the number of them. The problem is more combinatorial than ring theoretic.

Let m be an element in S. Then we can show by induction that

$$h^t(m) \equiv 2^t(m+1) - 1 \quad (\mathrm{mod}\, d).$$

Thus the order of the orbit containing m, the integer o_m, is the least integer $t > 0$ such that $2^t(m+1) \equiv m+1 \pmod{d}$. Let φ denote the Euler φ-function ([LeVeque, 1956] or any standard text on elementary number theory). Since d is odd, then $2^{\varphi(d)} \equiv 1 \pmod{d}$. Hence o_m divides $\varphi(d)$. In fact $o_m | o_0$. For if $2^{om}(m+1) \equiv (m+1) \pmod{d}$ then there is an integer u such that $(2^{om} - 1)(m+1) = ud$. Let s denote the greatest common divisor of $m+1$ and d and write $m+1 = \mu s$ and $d = \delta s$ with μ and δ relatively prime. Then $(2^{om} - 1)\mu = u\delta$. Hence $2^{om} - 1 \equiv 0 \pmod{\delta}$ and o_m is the order of 2 modulo δ. Also $2^{oo} \equiv 1 \pmod{\delta}$ so $o_m | o_0$.

If 2 is not a square modulo the prime d, then the orbit of h is all of S.

If $d \equiv 3 \pmod 6$, there is always an orbit of length 2 (and conversely). The orbit is $\{d - 3/3, 2d + 3/3\}$.

If $\mathfrak{o}$ has characteristic 2, then the ring $A = \mathfrak{o}[u, v, w]$ with $w^2 = u^i + v^j$ has $\mathrm{Cl}(A) = \mathrm{Cl}(\mathfrak{o}) \oplus \left(\coprod_{i=1}^{s} \mathbb{F}_{2^{o_i}} \cap \mathfrak{o} \right)$ (in the notation from above). □

Next we study power series rings. However the group of units of a power series ring and consequently the group of derived logarithms of units is much more complicated than the group of units of the polynomial ring. The next result is a step in the direction of finding the group of derived logarithms.

Lemma 17.9. *Suppose D is a derivation with the two properties.*

(a) *There are two elements x and y in the radical of B such that the ideal generated by $D(B)$ is generated by Dx and Dy and is not contained in a prime divisorial ideal (i.e., the elements Dx and Dy are relatively prime).*

(b) *There is an a in $\operatorname{Ker} D$ so that $D^2 = aD$ (in particular char $B = 2$). Then $V_0 = V_0' \cap D(B)B$.*

Proof. Suppose t is an element in B such that $t^{-1} Dt \in D(B)B$. By Lemma 17.6, the element $t^{-1} Dt$ satisfies an equation $D(t^{-1} Dt) + a(t^{-1} Dt) + (t^{-1} Dt)^2 = 0$. Suppose $t^{-1} Dt = r Dx + s Dy$. After expanding $D(t^{-1} Dt)$ and $(t^{-1} Dt)^2$ and substituting them into the equation and after simplifying the resulting expression we get the equation

$$Dx(Dr + r^2 Dx) = Dy(Ds + s^2 Dy).$$

Since Dx and Dy are relatively prime, there is an element b in B such that $Dr + r^2 Dx = b Dy$ and $Ds + r^2 Dy = b Dx$. Denote by u the element $1 + rx + sy + (rs + b)xy$. Then u is a unit in B and $u^{-1} Du = t^{-1} Dt$. □

Samuel [1964 b] asks whether it is always the case that $V_0 = V_0' \cap D(B)B$. Other special instances have been found where this holds. But there are also counterexamples (cf. [Hallier, 1964], [Yuan, 1968] and [Singh, 1968]).

The main application of this lemma is made in case $B = \mathfrak{o}[[X, Y]]$, the ring of power series in two indeterminates with coefficients in the Krull domain $\mathfrak{o}$. We assume the characteristic is 2. The derivation D will be defined so that the conditions of Lemma 17.9 and Proposition 17.5 are fulfilled. Then the kernel of the induced homomorphism $\operatorname{Cl}(A) \to \operatorname{Cl}(B)$ is isomorphic to V_0'/V_0. But $V_0 = V_0' \cap D(B)B$ so this group is isomorphic to $(V_0' + D(B)B)/D(B)B$, a group which we will be able to identify more precisely in terms of the other invariants of the extension.

Example 17.10. Let $\mathfrak{o}$ be a Krull domain of characteristic 2 and let B be the ring $\mathfrak{o}[[X, Y]]$. Define $D: B \to B$ by $DX = X$ and $DY = Y$ and extending to all of B. Then $A = \mathfrak{o}[[u, v, w]]$ with $w^2 = uv$ and where $u = X^2$, $v = Y^2$ and $w = XY$. It is immediately obvious that $D(B)B = BX + BY = BDX + BDY$. Hence the kernel of $i: \operatorname{Cl}(A) \to \operatorname{Cl}(B)$ is isomorphic to the group $V_0' + \mathfrak{m}/\mathfrak{m}$, where $\mathfrak{m} = BX + BY$. Since $D(X^r Y^s)$

$= (r+s) X^r Y^s$, this group is isomorphic to the cyclic group $\mathbb{Z}/2\mathbb{Z}$. Thus we get an exact sequence

$$0 \to \mathbb{Z}/2\mathbb{Z} \to \mathrm{Cl}(A) \to \mathrm{Cl}(B).$$

We would like to know the group $\mathrm{Cl}(B)$ in terms of $\mathrm{Cl}(\mathfrak{o})$. The most that can be generally said is that in case $\mathfrak{o}$ is noetherian there is a split monomorphism $\mathrm{Cl}(\mathfrak{o}) \to \mathrm{Cl}(B)$ by Theorem 18.8.

In fact we will use this descent technique later in this section in order to get an example for which $\mathrm{Cl}(\mathfrak{o}) \to \mathrm{Cl}(\mathfrak{o}[[T]])$ is not a bijection.

Example 17.11. Let $\mathfrak{o}$ and B be as above. Suppose i and j are two positive integers. Define $D : B \to B$ by setting $DX = Y^{2j}$ and $DY = X^{2i}$ and extending. Then $A = \mathfrak{o}[[u, v, w]]$ with $w^2 = u^{(2i+1)} + v^{(2j+1)}$. The ideal generated by $D(B)$ is generated by DX and DY. From Proposition 17.5, it follows that $\mathrm{Ker}\,(\mathrm{Cl}(A) \to \mathrm{Cl}(B)) = V_0'/V_0$.

Assign weights to monomials in X and Y by assigning the weight $2j+1$ to X and $2i+1$ to Y and extending. Let B_h denote the set of monomials of weight h together with 0. Now any power series can be written uniquely as an infinite sum $\sum_{h \geq 0} F_h$ with F_h in B_h. If $F = \sum_{h \geq q} F_h$ and $F_q \neq 0$, then the *order* of F is the integer q; in symbols $\mathrm{ord}\, F = q$. If F is a homogeneous polynomial of weight h, then DF is homogeneous of weight $h + 4ij - 1$.

Suppose F is an element in V_0', that is, there is an element u in L^* such that $u^{-1} Du = F$ and is an element in B. By Lemma 17.6, the element F must satisfy the equation $D^2 F - DF + F^2 = 0$. But $D^2 = 0$, so $F^2 = DF$. Suppose $q = \mathrm{ord}\, F$. Then $2q = \mathrm{ord}\, F^2$ and $\mathrm{ord}\, DF \geq q + 4ij - 1$. Hence $q \geq 4ij - 1$.

Filter V_0' by the subgroups W_h defined by $W_h = \{F \in V_0' : \mathrm{ord}\, F \geq h\}$. Denote by H the kernel of $\mathrm{Cl}(A) \to \mathrm{Cl}(B)$. Since $H = V_0'/V_0$, the group H is filtered by the subgroups $W_h + V_0/V_0$, which we denote by H_h. Denote by $\mathfrak{q}$ the ideal $D(B)B = B X^{2i} + B Y^{2j}$. Since $V_0 = V_0' \cap \mathfrak{q}$ and since $W_h \subseteq \mathfrak{q}$ if $h \geq 2(i+j)(i+j+1)$, it follows that $H_h = 0$ for sufficiently large n. Since the groups are vector spaces over the field $\mathbb{F}_2$, the extension problem is trivial, and hence $H = \coprod_{h \geq 4ij-1} H_h/H_{h+1}$, which is a finite sum. Since $\mathrm{ord}\, X^{2i} = 2i(2j+1)$ and $\mathrm{ord}\, Y^{2j} = 2j(2i+1)$, the group $V_0 \subseteq W_{4ij}$. Hence $H_{4ij-1}/H_{4ij} = W_{4ij-1}/W_{4ij}$. Now W_{4ij-1}/W_{4ij} is isomorphic to the kernel of $\mathrm{Cl}(\mathfrak{o}[X^2, Y^2, X^{2i+1} + Y^{2j+1}]) \to \mathrm{Cl}(\mathfrak{o}[X, Y])$, the group denoted by $P(2i+1, 2j+1)$ in Example 17.8. We have to determine the other part of the group H.

For each h, define $\varphi_h : W_h \to B_h$ by taking the component of weight h. Then $\ker \varphi_h = W_{h+1}$ for all h and $\varphi_h(W_h) = B_h \cap A$ for $h \geq 4ij$. As a consequence, $\varphi_h(W_h \cap \mathfrak{q}) = B_h \cap A \cap \mathfrak{q}$. In fact, to see that $\varphi_h(W_h) = B_h \cap A$, suppose $F = F_h + F_{h+1} + \cdots$ is a derived logarithm. Then $DF = F^2$ and weight $DF_h = h + 4ij - 1 < 2h$. But then $DF_h = 0$, so $F_h \in A$. If, on the other hand, the homogeneous $F_h \in B_h \cap A$, we must find an $F = F_h + F_{h+1} + \cdots$ such that $F^2 = DF$. If $F_h = 0$, then $F = 0$ works. Otherwise $DF = \sum_{q \geqslant h} DF_q$ with the weight DF_q equal to $q + 4ij - 1$. Now $DF = \sum_{q \geqslant h} F_q^2$ and the weight of each F_q^2 is $2q$. If q is even or $2q > h + 4ij - 1$, we must have $DF_q = 0$ while if q is odd we must have $q + 4ij - 1 = 2m$ with $DF_q = F_m^2$ for some F_m. In order to build such homogeneous elements we can do so for monomials $X^r Y^s$. That is, we find f such that $Df = (X^r Y^s)^2$ where $r(2j+1) + s(2i+1) \geqslant h \geqslant 4ij$. Then we can extend by linearity. Suppose $r(2j+1) + s(2i+1) \geqslant 4ij$. Then either $r \geqslant i$ or $s \geqslant j$. If $r \geqslant i$, let f be the element $X^{2(r-i)} Y^{2s+1}$. If $s \geqslant j$, then let f be the element $X^{2r+1} Y^{2(s-j)}$. In either case $Df = (X^r Y^s)^2$. Thus we can successfully integrate to find a solution to $DF = F^2$ with $\varphi_h(F)$ equal to the original F_h. As a consequence we get an isomorphism $H_h/H_{h+1} = B_h \cap A/B_h \cap A \cap \mathfrak{q}$ for $h \geqslant 4ij$. Therefore H_h/H_{h+1} is a free $\mathfrak{o}$-module of finite rank and hence H_{4ij}, which is just $\coprod_{h \geqslant 4ij} H_h/H_{h+1}$, is a free $\mathfrak{o}$-module of finite rank, say of rank $\mathrm{rk}(i, j)$. Now $\mathrm{rk}(i, j)$ is the number of monomials $X^{2r} Y^{2s}$ such that $0 \leqslant r < i$ and $0 \leqslant s < j$ with $r(2j+1) + s(2i+1) \geqslant 2ij$. Hence we have

$$H \cong P(2i+1, 2j+1) \oplus \mathfrak{o}^{\mathrm{rk}(i, j)}.$$

Samuel [1964b] remarks that $\mathrm{rk}(i, j)$ is asymptotic to $ij/2$. Also $\mathrm{rk}(i, j) = 0$ if and only if (i, j) is one of the pairs $(1, 1)$, $(1, 2)$ or $(2, 1)$. These give the ring $\mathfrak{o}[[u, v, w]]$ with $w^2 = u^3 + v^3$, the ring $\mathfrak{o}[[u, v, w]]$ with $w^2 = u^3 + v^5$, and the ring $\mathfrak{o}[[u, v, w]]$ with $w^2 = u^5 + v^3$ respectively. According to results of Scheja, these rings are factorial for sufficiently well behaved $\mathfrak{o}$ (cf. Section 19).

Proposition 17.12. *Let $\mathfrak{o}$ be a (noetherian) Krull domain of characteristic 2. Let (i, j) be nonzero integers with (i, j) distinct from $(1, 1)$, $(1, 2)$ and $(2, 1)$. Let A denote the ring $\mathfrak{o}[u, v, w]$ with $w^2 = u^{2i+1} + v^{2j+1}$ and denote by $\hat{A}$ the (u, v, w)-adic completion of A. Then $\mathrm{Cl}(A) \to \mathrm{Cl}(\hat{A})$ is not a bijection nor are the groups isomorphic.*

Proof. The class group $\mathrm{Cl}(A) = P(2i+1, 2j+1) \oplus \mathrm{Cl}(\mathfrak{o})$ while $\mathrm{Cl}(\hat{A})$ contains $P(2i+1, 2j+1) \oplus \mathfrak{o}^{\mathrm{rk}(i, j)}$ and a copy of $\mathrm{Cl}(\mathfrak{o})$. Moreover $\mathrm{rk}(i, j) \neq 0$. □

Using the same tricks, an example can be constructed of a (noetherian) Krull domain A for which $\mathrm{Cl}(A[[T]])$ is strictly bigger than $\mathrm{Cl}(A)$.

Example 17.13. Let $\mathfrak{o}$ be a Krull domain of characteristic 2. Denote by B the ring $\mathfrak{o}[X, Y]$ and by $\hat{B}$ the ring $\mathfrak{o}[[X, Y]]$. Let C be the ring of formal power series $B[[T]]$ with $\hat{C} = \hat{B}[[T]]$. Define derivations $D:C \to C$ and $\hat{D}:\hat{C} \to \hat{C}$ by setting $DX = Y^{2j}, DY = X^{2j}$ and $DT = 0$ and extending (resp: $DX = Y^{2j}$, etc.). Then Ker $D = \mathfrak{o}[X^2, Y^2, X^{2i+1} + Y^{2j+1}][[T]]$ and Ker $\hat{D} = \mathfrak{o}[[X^2, Y^2, X^{2i+1} + Y^{2j+1}]][[T]]$. As in the previous example we show that $V_0'(\hat{C})/V_0(\hat{C}) \cong V_0'(B)/V_0(B) \oplus (V_0'(\hat{C}) \cap TC)/(V_0(\hat{C}) \cap T\hat{C})$. Furthermore, the group $(V_0'(\hat{C}) \cap T\hat{C})/(V_0(\hat{C}) \cap T\hat{C})$ is isomorphic to countably many copies of the group $\mathfrak{o}^{\mathrm{rk}(i,j)}$. Thus in fact if $A = \mathfrak{o}[u, v, w]$ or $\mathfrak{o}[[u, v, w]]$, with $w^2 = u^{2i+1} + v^{2j+1}$, then $\mathrm{Cl}(A[[T]])/\mathrm{Cl}(A) \cong \mathfrak{o}[[T]]^{\mathrm{rk}(i,j)}$. Danilov [1970a] is able to explain, geometrically, the occurrance of this copy of $\mathfrak{o}[[T]]$.

Appendix

Waterhouse [1971] has combined the two theories of Galois and radical descent into one. The basic ingredient is the cohomology theory of Sweedler [1968].

Suppose A is a Krull domain with field of quotients K. Let B be the integral closure of A in a finite extension L of K. Suppose H is a cocommutative Hopf-A-algebra which admits B as an H-module algebra with fixed subring A. If D denotes the commutative dual Hopf algebra, $\mathrm{Hom}(H, A)$, then there is an algebra homomorphism $\sigma: B \to D \otimes_A B$ compatible with the Hopf structure of D. The extension $A \to B$ is said to be Galois if $(\sigma, 1 \otimes id): B \otimes_A B \to D \otimes_A B$ is an isomorphism and B is faithfully flat. It is said to be pseudo-Galois if it is Galois at each divisorial prime ideal of A.

Theorem A.1. *If B is a pseudo-Galois H-module algebra, then the canonical homomorphism* $\mathrm{Ker}(\mathrm{Cl}(A) \to \mathrm{Cl}(B)) \to \mathrm{H}^1(H, B)$ *is a bijection.*

For Galois descent, the Hopf algebra is the group ring $A[G]$ and then $\mathrm{H}^1(H, B) = \mathrm{H}^1(G, B^*)$. For radical descent for a derivation D, the Hopf algebra is $A[X]/(X^p - aX)$ and the $A[X]$-module algebra structure on B induces the derivation on B. More generally Yuan's [1968] theory is also contained in this generalization.

As an example, Waterhouse shows that $\mathrm{Cl}(k[x^q, y^q, xy])$ is cyclic of order q in case k is a factorial ring of characteristic p and q is a power of p.

The techniques of the proofs are very similar to those of Samuel reported in the two sections of this chapter.

Chapter V. Completions and Formal Power Series Extensions

This concluding chapter is concerned mainly with the behavior of the divisor class group under base change to the formal power series ring. It has been seen that $Cl(A) \to Cl(A[[T]])$ is not always a bijection, even when A is factorial. Using the good functorial properties of the Picard group and its relation to the divisor class group, Danilov has studied the class of rings for which the homomorphism is a bijection. A fundamental concept is the Picard group of an open subscheme of Spec A. For example, the set of points in Spec A at which an (divisorial) ideal is invertible is open, and on this subscheme, the class of the ideal belongs to the Picard group. Using this simple property, it is possible to show that there is a natural splitting of the injection $Cl(A) \to Cl(A[[T]])$.

The Picard group of the noetherian normal integral domain A is the subgroup of $Cl(A)$ generated by the classes of the invertible ideals. It is all of $Cl(A)$ if and only if A is locally factorial, which is explained by the fact that a local ring has a trivial Picard group. How then does the Picard group help to study the group $Cl(A)$ when A is local? In this case we consider the subscheme Spec $A - \{m\}$ and its Picard group. For example, if this Picard group is zero, then A has many fine properties.

The basic properties of the Picard group and its relations to the divisor class group are the main topics in Section 18.

In Section 19 we discuss Danilov's results concerning rings A for which $Cl(A) \to Cl(A[[T]])$ is a bijection. We begin with Scheja's result: If A is a complete local noetherian factorial ring with depth $A \geqslant 3$, then $Cl(A) \to Cl(A[[T]])$ is a bijection. If depth $A \leqslant 1$, then A is a regular local ring, so the map is a bijection by results in Section 18. Thus the main problems arise in case depth $A = 2$ or in case A is not complete.

Danilov's methods employ difficult results in algebraic geometry; for example some of the results concerning the resolutions of singularities due to Hironaka.

§ 18. The Picard Group

Suppose $(X, \mathcal{O}_X)$ is a ringed space. The set of isomorphism classes of invertible $\mathcal{O}_X$-modules is a group isomorphic to $H^1(X, \mathcal{O}_X^*)$ and is called the Picard group of X. The group induces a contravariant functor Pic from the category of ringed spaces to the category of abelian groups. The main purpose of this section is to introduce the theory of the Picard group and especially its applications to the study of the divisor class group. That the Picard groups of the open subschemes of Spec A can tell us much about Cl(A) is seen in the study of parafactorial rings. Using the properties of parafactorial rings, Grothendieck has shown that a local complete intersection which is factorial in codimension 3 is factorial. One can use this result to prove the Klein-Nagata theorem which we have established, using different methods, in Section 11. Now $\text{Pic}(A) \to \text{Pic}(A[[T]])$ is always a bijection. So if A and $A[[T]]$ are locally factorial (for example if A is a regular ring), then $\text{Cl}(A) \to \text{Cl}(A[[T]])$ is a bijection. Thus we get Claborn's theorem: If A is a regular ring, then $\text{Cl}(A) \to \text{Cl}(A[[T]])$ is a bijection.

Let $(X, \mathcal{O}_X)$ be a ringed space with $\mathcal{O}_X$ a sheaf of commutative rings. An $\mathcal{O}_X$-Module $\mathscr{L}$ is invertible if $\mathscr{L}$ is locally free of rank 1. If $\mathscr{L}$ is invertible, then the dual $\check{\mathscr{L}} = \mathscr{H}om_{\mathcal{O}_X}(\mathscr{L}, \mathcal{O}_X)$ is also invertible and there are isomorphisms

$$\mathscr{L} \otimes_{\mathcal{O}_X} \check{\mathscr{L}} \cong \mathscr{H}om_{\mathcal{O}_X}(\mathscr{L}, \mathscr{L}) \cong \mathcal{O}_X .$$

Denote by $\mathscr{P}$ the set of isomorphism classes of invertible $\mathcal{O}_X$-Modules. The set $\mathscr{P}$ has a group structure induced by tensor product, with inverse given by the class of the dual. It is not difficult to see that this group is isomorphic to the first cohomology group $H^1(X, \mathcal{O}_X^*)$ [Grothendieck and Dieudonné, EGA, 0_1].

This group is called the *Picard group* of $(X, \mathcal{O}_X)$ and is denoted by Pic(X). If $(X, \mathcal{O}_X) = (\text{Spec } A, \tilde{A})$ which is the affine scheme associated to the commutative ring A, then we write Pic(A) in place of Pic(Spec A).

For each open set U in X, let S_U denote the set of elements in $\Gamma(U, \mathcal{O}_X)$ which are locally regular, so that $S_U = \{s \in \Gamma(U, \mathcal{O}_X) : s_x$ is regular in $\mathcal{O}_{X,x}$ for all x in $U\}$. Then S_U is a multiplicatively closed subset of $\Gamma(U, \mathcal{O}_X)$. The assignments $U \to S_U$ define a presheaf of sets on X. Hence the assignments $U \to S_U^{-1}\Gamma(U, \mathcal{O}_X)$ for each open subset U in X define a presheaf on X which is in fact a sheaf, which we denote by $\mathscr{M}_X$. The sheaf $\mathcal{O}_X$ is canonically a subsheaf of $\mathscr{M}_X$ and $\mathcal{O}_X^*$ is canonically a subsheaf of $\mathscr{M}_X^*$. There is then an exact sequence of abelian sheaves on X

$$1 \to \mathcal{O}_X^* \to \mathscr{M}_X^* \to \mathscr{M}_X^*/\mathcal{O}_X^* \to 0 .$$

There is induced a long exact sequence of cohomology groups

$$1 \to \Gamma(X, \mathcal{O}_X^*) \to \Gamma(X, \mathcal{M}_X^*) \to \Gamma(X, \mathcal{M}_X^*/\mathcal{O}_X^*) \to \mathrm{H}^1(X, \mathcal{O}_X^*) \to \mathrm{H}^1(X, \mathcal{M}_X^*) \to \cdots.$$

The group $\Gamma(X, \mathcal{M}_X^*/\mathcal{O}_X^*)$ is called the group of *Cartier divisors* on X and is denoted by $\mathrm{Cart}(X)$. (In this same sense the group $\mathrm{Div}(A)$ should be called the group of Weil divisors on A.)

Suppose X is the affine scheme $\mathrm{Spec}\, A$. Then $\Gamma(X, \mathcal{M}_X)$ is the total ring of quotients of A, say K, and the long exact sequence becomes

$$1 \to A^* \to K^* \to K^*/A^* \to \mathrm{Pic}\, A \to \mathrm{H}^1(X, \mathcal{M}_X^*) \to \cdots.$$

An invertible $\mathcal{O}_X$-Module is then sheaf associated to an invertible A-module.

The Cartier divisors of X can be described by fractionary $\mathcal{O}_X$-Ideals. A fractionary $\mathcal{O}_X$-Ideal is an $\mathcal{O}_X$-submodule of $\mathcal{M}_X$. It is invertible if it is invertible as an $\mathcal{O}_X$-Module. Let $\mathrm{Inv}(X)$ denote the group of invertible fractionary $\mathcal{O}_X$-Ideals. The presheaves defined by the assignments $U \mapsto \mathrm{Cart}\, U$ and $U \mapsto \mathrm{Inv}(U)$ are sheaves on X denoted by Cart_X and Inv_X respectively.

Lemma 18.1. *A fractionary $\mathcal{O}_X$-Ideal $\mathcal{I}$ is invertible if and only if for each x in X, there is an open subset U of X containing x and an f in $\Gamma(U, \mathcal{M}_X^*)$ such that $\mathcal{I}_U \cong \mathcal{O}_U \cdot f$.*

Proof. [EGA IV, 21.2.2., 1967] □

Thus, for each open subset U of X there is induced a morphism $\Gamma(U, \mathcal{M}_X^*) \to \mathrm{Inv}\, U$ and consequently an isomorphism $\mathrm{Cart}_X \to \mathrm{Inv}_X$ of sheaves of X. The corresponding composition $\mathrm{Cart}(X) \to \mathrm{Inv}(X) \to \mathrm{Pic}(X)$ is just the natural homomorphism $\mathrm{Cart}(X) \to \mathrm{H}^1(X, \mathcal{O}_X^*)$.

Suppose A is an integral domain with field of quotients K and P is an invertible A-module. Then $K \otimes_A P$ is a one dimensional vector space over K and consequently isomorphic to K. The natural map $P \to K \otimes_A P$ is an injection. Hence P is isomorphic to an invertible fractionary A-ideal in K. In general, it is not the case that invertible A-modules are isomorphic to ideals (see [Garfinkel, 1971] for example). One can ask when $\mathrm{Cart}\, X \to \mathrm{Pic}\, X$ is a surjection.

Proposition 18.2. *Let X be a scheme which is either* (a) *locally noetherian with $\mathrm{Ass}\, X$ contained in an open affine or,* (b) *reduced with a locally finite set of irreducible components. Then every invertible $\mathcal{O}_X$-Module is isomorphic to an invertible fractionary $\mathcal{O}_X$-Ideal. In particular $\mathrm{Cart}(X) \to \mathrm{Pic}(X)$ is a surjection.*

Proof. [EGA, IV, 21.3.4, 1967] ⬚

Corollary 18.3. *Suppose A is a Krull domain. Then there is a natural injection* $\operatorname{Pic} A \to \operatorname{Cl}(A)$.

Proof. In fact $\operatorname{Cart}(A)$ is a subgroup of $\operatorname{Div}(A)$ which contains $\operatorname{Prin}(A)$. ⬚

Suppose $X \to X'$ is a morphism of ringed spaces. There is induced a group homomorphism $\operatorname{Pic} X' \to \operatorname{Pic} X$ by $\mathscr{L}' \mapsto \mathscr{L} \otimes_{\mathcal{O}_{X'}} \mathcal{O}_X$. Thus $X \mapsto \operatorname{Pic} X$ is a contravariant functor from the category of ringed spaces to the category of abelian groups.

In case A is a Krull domain, the Corollary above shows that $\operatorname{Pic} A$ is a subgroup of $\operatorname{Cl}(A)$. If $A \to A'$ is a ring homomorphism, then there is induced a homomorphism $\operatorname{Pic} A \to \operatorname{Pic} A'$. If $\operatorname{Pic} A = \operatorname{Cl}(A)$, then this induced homomorphism can be used to study $\operatorname{Cl}(A)$.

Suppose X is a locally noetherian scheme. Let $X^{(1)}$ denote the set $\{x \in X : \dim \mathcal{O}_{X,x} = 1\}$. Let $\operatorname{Cyc}^1(X)$ denote the subgroup of the product $\mathbb{Z}^{X^{(1)}}$ consisting of those functions $f : X^{(1)} \to \mathbb{Z}$ which have locally finite support. In case $X = \operatorname{Spec} A$ with A a normal noetherian integral domain, then $\operatorname{Cyc}^1(X)$ is our old friend $\operatorname{Div}(A)$.

(It is possible, using the theory of 1-noetherian rings developed in [Claborn and Fossum, 1968], to define locally 1-noetherian schemes. Then one could define the group of 1-cycles. There is a reason to do this, for then the groups $\operatorname{Cyc}^1(X)$ and $\operatorname{Cl}(X)$, for a locally Krull scheme X, can be defined. Since we make no use of this here, since the main applications are to noetherian normal domains, there is no need to delve into this theory at this point.)

For the locally noetherian scheme X, the group $\operatorname{Cyc}^1(X)$ will, from now on, be denoted by $\operatorname{Div}(X)$. For each open subset U in X, the group $\operatorname{Div}(U)$ is defined. The assignment $U \mapsto \operatorname{Div}(U)$ defines a flabby sheaf on X denoted by Div_X.

There is a canonical homomorphism $\mathscr{M}_X^* \to \operatorname{Div}_X$ whose kernel is $\mathcal{O}_X^*$. To define this, it is sufficient to define a homomorphism $\Gamma(U, \mathscr{M}_X^*) \to \operatorname{Div}(U)$ for each open set U in X which is compatible with the restrictions. But $\Gamma(U, \mathscr{M}_X^*)$ is just $(S_U^{-1} \mathcal{O}_X(U))^*$. A unit u in $S_U^{-1} \mathcal{O}_X(U)$ is a fraction f/s where both f and s are locally regular elements in $\mathcal{O}_X(U)$. If $x \in X^{(1)} \cap U$ and f is locally regular in $\mathcal{O}_X(U)$, then the $\mathcal{O}_{X,x}$-module $\mathcal{O}_{X,x} / f \mathcal{O}_{X,x}$ has finite length. Define the function $h_u : U^{(1)} \to \mathbb{Z}$ by $h_u(x) = \operatorname{length}(\mathcal{O}_{X,x}/f\mathcal{O}_{X,x}) - \operatorname{length}(\mathcal{O}_{X,x}/s\mathcal{O}_{X,x})$. It is easy to verify that $h_u \in \operatorname{Div}(U)$. The assignments $u \mapsto h_u$ define a homomorphism $(S_U^{-1}\mathcal{O}_X(U))^* \to \operatorname{Div} U$ which is compatible with the restriction maps. Hence there is induced a sheaf homomorphism $\mathscr{M}_X^* \to \operatorname{Div}_X$ which clearly vanishes on $\mathcal{O}_X^*$. Hence there is induced a morphism $\operatorname{Cart}_X \to \operatorname{Div}_X$.

Denote the cokernel of the homomorphism $\Gamma(X, \mathcal{M}_X^*) \to \Gamma(X, \mathrm{Div}_X)$ by $\mathrm{Cl}(X)$. We have the commutative diagram

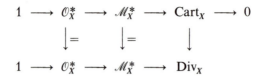

with exact rows which yields, upon application of the global section functor, the commutative diagram

$$1 \longrightarrow \Gamma(X, \mathcal{O}_X^*) \longrightarrow \Gamma(X, \mathcal{M}_X^*) \longrightarrow \mathrm{Cart}(X) \longrightarrow \mathrm{Pic}(X) \ldots$$

$$1 \longrightarrow \Gamma(X, \mathcal{O}_X^*) \longrightarrow \Gamma(X, \mathcal{M}_X^*) \longrightarrow \mathrm{Div}(X) \longrightarrow \mathrm{Cl}(X) \longrightarrow 0.$$

We ask the following questions: When is there induced a homomorphism $\mathrm{Pic}(X) \to \mathrm{Cl}(X)$? When is it a monomorphism? When is the homomorphism $\mathrm{Cart}(X) \to \mathrm{Div}(X)$ a monomorphism? If A is a normal noetherian integral domain and $X = \mathrm{Spec}\, A$, then $\mathrm{Cart}(X) \to \mathrm{Pic}(X)$ is a surjection, so there is induced a homomorphism $\mathrm{Pic}(X) \to \mathrm{Cl}(X)$ which is a monomorphism. Slightly more general is the next result.

Proposition 18.4. *Suppose X is a normal locally noetherian scheme. Then the canonical homomorphism $\mathrm{Cart}(X) \to \mathrm{Div}(X)$ is an injection. The homomorphism $\mathrm{Cart}(X) \to \mathrm{Pic}(X)$ is a surjection. The homomorphism $\mathrm{Cart}(X) \to \mathrm{Div}(X)$ is a bijection if and only if each local ring $\mathcal{O}_{X,x}$ is factorial.*

Proof. [EGA, IV, 21.6.9, 1967] □

Corollary 18.5. *Let X be a locally noetherian normal scheme. Then $\mathrm{Pic}(X) \to \mathrm{Cl}(X)$ is an injection. It is a bijection if and only if $\mathcal{O}_{X,x}$ is factorial for each x in X.* □

Suppose A is a normal noetherian integral domain which has only a finite number of maximal ideals at which it is locally not factorial, say $\mathfrak{m}_1, \ldots, \mathfrak{m}_n$. Let S be the multiplicatively closed subset $A - \bigcup\limits_{i=1}^{n} \mathfrak{m}_i$. Then $S^{-1}A$ is a semi-local ring. We can calculate $\mathrm{Cl}(A)$ in terms of $\mathrm{Cl}(S^{-1}A)$ and $\mathrm{Pic}(A)$.

Corollary 18.6. *Let $\mathfrak{m}_1, \ldots, \mathfrak{m}_n$ be the maximal ideals of A at which $A_{\mathfrak{m}_i}$ is not factorial. Then $\mathrm{Pic}(A)$ is the kernel of the surjection $\mathrm{Cl}(A) \to \mathrm{Cl}(S^{-1}A)$, the set S being the complement of the union of the $\mathfrak{m}_j$. That is, the sequence $0 \to \mathrm{Pic}(A) \to \mathrm{Cl}(A) \to \mathrm{Cl}(S^{-1}A) \to 0$ is exact.*

Proof. Suppose $\mathfrak{a}$ is a divisorial ideal whose class is in the kernel of $\mathrm{Cl}(A) \to \mathrm{Cl}(S^{-1}A)$. Then $\mathfrak{a}A_{\mathfrak{m}_i}$ is principal for each i, while $\mathfrak{a}A_{\mathfrak{m}}$ is principal for each maximal ideal $\mathfrak{m}$ distinct from the $\mathfrak{m}_j$. Hence $\mathfrak{a}$ is locally invertible and so invertible. Thus the class of $\mathfrak{a}$ is in $\mathrm{Pic}(A)$. Since $S^{-1}A$ is semi-local, invertible ideals in $S^{-1}A$ are principal. Thus we get our exact sequence

$$0 \to \mathrm{Pic}(A) \to \mathrm{Cl}(A) \to \mathrm{Cl}(S^{-1}A) \to 0 \, . \quad \square$$

Next the class group $\mathrm{Cl}(X)$ will be compared with the Picard group of open subschemes.

Proposition 18.7. *Suppose X is a normal noetherian scheme and that $\{U_c\}_{c \in C}$ is a filtered family of open subsets of X. Consider the conditions:*
    (a) *For all c in C and all x in $X - U_c$, the Krull dimension $\dim \mathcal{O}_{X,x} \geqslant 2$.*
    (b) *For all x in $\bigcap_{c \in C} U_c$, the local ring $\mathcal{O}_{X,x}$ is factorial.*
If (a) is satisfied, then the homomorphisms $\mathcal{O}_X(X)^* \to \mathcal{O}_X(U_c)^*$, $\mathcal{M}_X(X)^* \to \mathcal{M}_X(U_c)^*$, $\mathrm{Div}(X) \to \mathrm{Div}(U_c)$ and $\mathrm{Cl}(X) \to \mathrm{Cl}(U_c)$ are bijections for all c in C. In particular, for each c in C there is induced an injection $\mathrm{Pic}(U_c) \to \mathrm{Cl}(X)$.
If, in addition, condition (b) is satisfied, then $\varinjlim \mathrm{Pic}(U_c) \to \mathrm{Cl}(X)$ is a bijection.

Proof [EGA, IV, 21.6.12, 1967]. If the U_c satisfy condition (a), then $X^{(1)} \subseteq \bigcap_{c \in C} U_c$. Hence $\mathrm{Div}(X) = \mathrm{Div}(U_c)$ for all c. The remainder of the proof is rather straightforward. $\square$

This result has been used effectively by Danilov [1970b] to compare $\mathrm{Cl}(A)$ and $\mathrm{Cl}(A[[T]])$. Denote by i^* the natural homomorphism $i^*: \mathrm{Cl}(A) \to \mathrm{Cl}(A[[T]])$.

Theorem 18.8. *Let A be a normal noetherian integral domain. The homomorphism $j: A[[T]] \to A$ given by $j(f(T)) = f(0)$ for each element $f(T)$ in $A[[T]]$ induces a homomorphism $j^*: \mathrm{Cl}(A[[T]]) \to \mathrm{Cl}(A)$ such that $j^* \cdot i^* = \mathrm{id}$. In particular i^* is an injection and j^* is a surjection.*

Proof. Let X and Y denote respectively $\mathrm{Spec}\, A$ and $\mathrm{Spec}\, A[[T]]$. Then j induces a closed regular immersion $X \to Y$ which identifies X with $V(T)$. Let $\mathfrak{a}$ be a divisorial ideal in $A[[T]]$. The set $\{y \in Y : \mathfrak{a}_y$ is invertible$\}$ is an open set in Y which contains all the regular points in Y and in particular the divisorial prime ideals. Let V denote this set. Then $X \cap V$ is an open subset of X which contains all the regular points of X. In particular $X^{(1)} \subseteq X \cap V$ since A is normal. Since the sheaf of ideals $\tilde{\mathfrak{a}}|_V$ is invertible on V, its induced sheaf $j^*(\tilde{\mathfrak{a}}|_V)$ is an invertible sheaf of ideals on $X \cap V$. Therefore it defines an element in $\mathrm{Pic}(X \cap V)$. By

Proposition 18.7, the group $\text{Pic}(X \cap V)$ is canonically a subgroup of $\text{Cl}(A)$. Thus we get an assignment $\mathfrak{a} \mapsto \text{cl}(j^*(\tilde{\mathfrak{a}}|_V))$ which is obviously independent of the class of $\mathfrak{a}$, and which induces a homomorphism $j^*: \text{Cl}(A[[T]]) \to \text{Cl}(A)$. It is clear that $j^*(\text{cl}(\mathfrak{a} A[[T]])) = \text{cl}(\mathfrak{a})$ for each divisorial ideal $\mathfrak{a}$ in A. $\square$

Using the same idea, we can show that $\text{Cl}(A)$ is covered by $\text{Pic}(U)$ for open subsets of X. For a divisorial ideal $\mathfrak{a}$ in A, let $U_\mathfrak{a} = \{x \in \text{Spec } A : \mathfrak{a} A_x$ is invertible$\}$. Then $U_\mathfrak{a}$ is open since $U_\mathfrak{a} = \{x \in \text{Spec } A : (\mathfrak{a} \otimes_A \text{Hom}_A(\mathfrak{a}, A))_x \to \text{Hom}_A(\mathfrak{a}, \mathfrak{a})_x$ is a surjection$\}$. The family $\{U_\mathfrak{a}\}_{\mathfrak{a} \subseteq A}$ is filtered by inclusion and each $U_\mathfrak{a}$ contains all the regular points of $\text{Spec } A$.

Proposition 18.9. *The induced homomorphism* $\varinjlim \text{Pic}(U_\mathfrak{a}) \to \text{Cl}(A)$ *is a bijection.*

Proof. If $x \in \bigcap_\mathfrak{a} U_\mathfrak{a}$, then A_x is factorial since every divisorial ideal is principal. The result follows immediately from Proposition 18.7. $\square$

Suppose A is a noetherian local ring which is normal. Let $U = \text{Spec } A - \{\mathfrak{m}\}$, where $\mathfrak{m}$ is the maximal ideal. Then Proposition 18.7 can be used to examine $\text{Cl}(A)$.

Proposition 18.10. *Suppose A is a normal local ring with* $\dim A \geqslant 2$.
(a) $\text{Pic}(U) \to \text{Cl}(A)$ *is an injection.*
(b) *If $A_\mathfrak{p}$ is factorial for all $\mathfrak{p}$ in U, then* $\text{Pic}(U) \to \text{Cl}(A)$ *is a bijection.*
(c) *The ring A is factorial if and only if* $\text{Pic}(U) = 0$ *and $A_\mathfrak{p}$ is factorial for all $\mathfrak{p}$ in U.*

Proof. This follows immediately from Proposition 18.7. $\square$

A local noetherian ring is said to be *parafactorial* if depth $A \geqslant 2$ and $\text{Pic}(U) = 0$. If A is normal, then depth $A \geqslant 2$ by Theorem 4.1.

Corollary 18.11. *Suppose A is a noetherian local ring with* $\dim A \geqslant 2$. *Then A is factorial if and only if A is parafactorial and $A_\mathfrak{p}$ is factorial for all $\mathfrak{p}$ in U.*

Proof. It is enough to show that A is normal if it is parafactorial and locally factorial at elements in U. (The other direction is part (c) of the proposition.)
But A is normal according to Theorem 4.1. For if $\mathfrak{p} \in \text{Spec } A$ with $\text{ht } \mathfrak{p} = 1$, then $\mathfrak{p} \neq \mathfrak{m}$, and so $A_\mathfrak{p}$ is factorial. Hence $A_\mathfrak{p}$ is a principal valuation ring. Otherwise we have depth $A_\mathfrak{p} \geqslant \inf(2, \text{ht } \mathfrak{p})$ for $\mathfrak{p}$ in U and this holds for $\mathfrak{p}$ replaced by $\mathfrak{m}$ as well since A is parafactorial. $\square$

The main reason for introducing the notion of parafactorial rings is to give an outline of the proof of the Grothendieck-Samuel Theorem.

Recall that the noetherian local ring A is a *complete intersection* if the completion $\hat{A}$ is the factor of a regular local ring by an ideal generated by a regular sequence.

Proposition 18.12. *Suppose A is a local ring. If the completion $\hat{A}$ is parafactorial, then A is parafactorial.*

Proof. [Grothendieck, XI, Prop. 3, 1963] ☐

Proposition 18.13. *Suppose A is a complete intersection. If $\dim A \geqslant 4$, then A is parafactorial.*

Proof. It is enough to show that the completion A is parafactorial. Suppose B is a regular local ring such that $A = B/(f_1, \ldots, f_r)$ where $f_1, \ldots, f_r$ is a regular B-sequence. Then $\dim A = \dim B - r$. We may assume B is complete. Since B is regular, it is factorial and parafactorial. The proof is concluded by induction on the Krull dimension, the inductive bridge being contained in the following lemma. ☐

Suppose t is a regular element in the maximal ideal $\mathfrak{m}$ of the local ring A. Denote by A' the residue ring A/tA. Let X_1 denote $\operatorname{Spec} A$ and X_1' denote $\operatorname{Spec} A'$. Then $X_1' = V(t)$ is a closed subscheme of X_1. Denote by X the open subscheme $\operatorname{Spec} A - \{\mathfrak{m}\}$ of X_1. The open subscheme $\operatorname{Spec} A' - \{\mathfrak{m}'\}$ is to be denoted by X'. So we have

$$X_1 = \operatorname{Spec} A, \qquad X_1' = \operatorname{Spec} A' = V(t),$$
$$X = X_1 - \{\mathfrak{m}\} \quad \text{and} \quad X' = X_1' \cap X.$$

The immersion $X' \to X$ induces a homomorphism $\operatorname{Pic}(X) \to \operatorname{Pic}(X')$. The lemma tells when this is an isomorphism (and completes the induction bridge for the preceding theorem).

Consider the following conditions:

(a) *For each closed point x in X, the local ring $\mathcal{O}_{X,x}$ has $\operatorname{depth} \mathcal{O}_{X,x} \geqslant 2$.* (That is, $\operatorname{depth} A_{\mathfrak{p}} \geqslant 2$ for each prime ideal $\mathfrak{p}$ with $\dim A/\mathfrak{p} = 1$.)

(b) *The ring A/tA has $\operatorname{depth} A/tA \geqslant 3$.*

(b') *The ring A/tA has $\operatorname{depth} A/tA \geqslant 4$.*

(c) *For each closed point y in X', the ring $\mathcal{O}_{X,y}$ has $\operatorname{depth} \mathcal{O}_{X,y} \geqslant 3$.* (That is for each $\mathfrak{p}$ in $V(t)$ such that $\dim A/\mathfrak{p} = 1$, $\operatorname{depth} A_{\mathfrak{p}} \geqslant 3$.)

(d) *For each closed point x in $X - X'$, the ring $\mathcal{O}_{X,x}$ is parafactorial.* (If $\dim A/\mathfrak{p} = 1$ and $t \notin \mathfrak{p}$, then $A_{\mathfrak{p}}$ is parafactorial.)

Lemma 18.14. *Suppose A is complete in the (t)-adic topology.*

(i) *Then (a) and (b) imply that $\operatorname{Pic}(X) \to \operatorname{Pic}(X')$ is an injection.*

(ii) *Conditions (a), (b'), (c) and (d) imply that $\operatorname{Pic}(X) \to \operatorname{Pic}(X')$ is a bijection. In particular A' is parafactorial if and only if A is parafactorial.*

Proof. [Grothendieck, XI, Lemmata 3.16 and 3.17, 1962] ☐

The proof of this lemma is very geometric. The Theorem 18.13 has a very nice application to the theory of factorial rings, the reason for its inclusion in this section.

Proposition 18.15. *Let A be a complete intersection. If $A_\mathfrak{p}$ is factorial for all $\mathfrak{p}$ with $\operatorname{ht} \mathfrak{p} \geqslant 3$, then A is factorial.*

Proof. We go by induction on $\dim A$. If $\dim A \leqslant 3$, then A is factorial by assumption. If $\dim A > 3$, then it is sufficient, by Proposition 18.11, to show that $A_\mathfrak{p}$ is factorial for all non-maximal prime ideals $\mathfrak{p}$.

But $A_\mathfrak{p}$ is again a complete intersection which is factorial in codimension 3. Furthermore $\dim A_\mathfrak{p} < \dim A$. Hence $A_\mathfrak{p}$ is factorial. ∎

The next lemma is a very simple analogue of Lemma 18.13. But it is very useful.

Lemma 18.16. *Let $\mathfrak{a}$ be an ideal contained in the Jacobson radical of A. Then the natural homomorphism $\operatorname{Pic}(A) \to \operatorname{Pic}(A/\mathfrak{a})$ is an injection. If A is complete in the $\mathfrak{a}$-adic topology, then it is a bijection.*

Proof. If L is an invertible A-module such that $L/\mathfrak{a}L$ is free as an $A/\mathfrak{a}$-module, then an element in L which generates $L/\mathfrak{a}L$ also generates L.

Suppose that A is complete. Let $\overline{L}$ be an invertible (or projective of finite type) $A/\mathfrak{a}$-module. Suppose $\overline{L} \oplus \overline{L}_1$ is free of rank n. Let Λ be the endomorphism ring of A^n. Then $\Lambda/\mathfrak{a}\Lambda$ is isomorphic to the endomorphism ring of $\overline{L} \oplus \overline{L}_1$. Since A is complete in the $\mathfrak{a}$-adic topology, idempotents in $\Lambda/\mathfrak{a}\Lambda$ can be lifted to idempotents in Λ. Let e be an idempotent in Λ whose image in $\Lambda/\mathfrak{a}\Lambda$ has $\overline{L}$ as kernel. Then $\ker e$ is an invertible (projective) A-module such that $\ker e/\mathfrak{a}(\ker e) \cong \overline{L}$. Hence $\operatorname{Pic}(A) \to \operatorname{Pic}(A/\mathfrak{a})$ is bijective. ∎

Corollary 18.17. *Suppose $\mathfrak{a}$ is an ideal contained in the radical of A. Let $\hat{A}$ denote the $\mathfrak{a}$-adic completion of A. Then $\operatorname{Pic}(A) \to \operatorname{Pic}(\hat{A})$ is an injection.*

Proof. Consider the commutative diagram

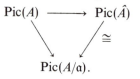

Now $\operatorname{Pic}(A) \to \operatorname{Pic}(A/\mathfrak{a})$ and $\operatorname{Pic}(\hat{A}) \to \operatorname{Pic}(A/\mathfrak{a})$ are both injections. Hence also $\operatorname{Pic}(A) \to \operatorname{Pic}(\hat{A})$ is an injection. ∎

Recall that the noetherian ring A is *regular* if $A_\mathfrak{p}$ is a regular local ring for all $\mathfrak{p}$ in $\operatorname{Spec} A$. Suppose $\hat{A}$ denotes the $\mathfrak{a}$-adic completion of A.

Then $\hat{A}$ is regular if and only if $A_{\mathfrak{m}}$ is regular for all prime (maximal) ideals $\mathfrak{m}$ in $V(\mathfrak{a})$ [Greco and Salmon, Proposition 8.3, 1971]. If $\mathfrak{a}$ is an ideal in the radical of A, then $\hat{A}$ is regular if and only if A is regular.

Proposition 18.18. *Suppose* $\hat{A}$ *is locally factorial and* $\operatorname{Spec} A/\mathfrak{a}$ *is connected (i.e., $A/\mathfrak{a}$ has no nontrivial idempotents). Then* $\hat{A}$ *is a normal integral domain with* $\operatorname{Cl}(\hat{A})$ *naturally isomorphic to* $\operatorname{Pic}(A/\mathfrak{a})$. *That is* $\operatorname{Pic}(A/\mathfrak{a}) = \operatorname{Pic}(\hat{A})$ *and* $\operatorname{Pic}(\hat{A}) \to \operatorname{Cl}(\hat{A})$ *is a bijection.*

Proof. Since A is locally an integral domain and $\operatorname{Spec} A/\mathfrak{a}$ is connected, the completion $\hat{A}$ is an integral domain [Greco and Salmon, Proposition 10.8, 1971]. It is then normal by Theorem 4.1. Since it is locally factorial, the inclusion $\operatorname{Pic}(\hat{A}) \to \operatorname{Cl}(\hat{A})$ is a bijection. Then $\operatorname{Pic}(\hat{A}) = \operatorname{Pic}(A/\mathfrak{a})$ by Lemma 18.16. $\square$

Corollary 18.19. *The ring* $\hat{A}$ *is factorial if and only if* $\operatorname{Pic}(A/\mathfrak{a}) = 0$. $\square$

Corollary 18.20. *Suppose* $\mathfrak{a}$ *is an ideal in A such that* $\operatorname{Spec} A/\mathfrak{a}$ *is connected and the elements in $V(\mathfrak{a})$ are regular (i.e., $A_{\mathfrak{p}}$ is regular for all $\mathfrak{p}$ in $V(\mathfrak{a})$). Then* $\operatorname{Cl}(\hat{A}) = \operatorname{Pic}(A/\mathfrak{a})$. $\square$

This corollary leads to the construction of factorial rings whose completions are not factorial. For example, set $A = \mathbb{R}[X, Y]$. Then $A/(X^2 + Y^2 - 1)$ is a regular integral domain with $\operatorname{Cl}(A/(X^2 + Y^2 - 1)) = \mathbb{Z}/2\mathbb{Z}$, which is also $\operatorname{Pic}(A/(X^2 + Y^2 - 1))$. Therefore the $(X^2 + Y^2 - 1)$-adic completion of A is a regular integral domain with divisor class group isomorphic to $\mathbb{Z}/2\mathbb{Z}$.

Proposition 18.21. *The extension* $A \to A[[T]]$ *induces a bijection* $\operatorname{Pic}(A) \to \operatorname{Pic}([[T]])$.

Proof. Since T is in the radical of $A[[T]]$ and since $A[[T]]$ is (T)-adically complete, the morphism $\operatorname{Pic}(A[[T]]) \to \operatorname{Pic}(A)$, induced by $T \mapsto 0$, is a bijection by Lemma 18.16. The composition $\operatorname{Pic}(A) \to \operatorname{Pic}(A[[T]]) \to \operatorname{Pic}(A)$ is the identity. $\square$

We can apply these results concerning the Picard group together with Mori's theorem (Corollary 6.12) to obtain the following result due to Samuel [1961a].

Proposition 18.22. *Suppose A is a normal noetherian integral domain. If the local rings $A_{\mathfrak{m}}[[T]]$ are factorial for all maximal ideals $\mathfrak{m}$ of A, then both A and $A[[T]]$ are locally factorial and in particular* $\operatorname{Cl}(A) \to \operatorname{Cl}(A[[T]])$ *is a bijection.*

Proof. By Corollary 6.11 the homomorphism $\operatorname{Cl}(A_{\mathfrak{m}}) \to \operatorname{Cl}(A_{\mathfrak{m}}[[T]])$ is an injection. Therefore A is locally factorial. Let $\mathfrak{M}$ be a maximal ideal of $A[[T]]$ and denote by $\mathfrak{m}$ the maximal ideal $\mathfrak{M} \cap A$. Then

$\mathfrak{M} = \mathfrak{m} A[[T]] + (T)$. The pair $(A[[T]]_{\mathfrak{M}}, (T))$ is a Zariski-pair whose completion is $A_{\mathfrak{m}}[[T]]$. If $A_{\mathfrak{m}}[[T]]$ is factorial, then so is $A[[T]]_{\mathfrak{M}}$. Hence $A[[T]]$ is locally factorial. Thus $\mathrm{Pic} \to \mathrm{Cl}$ is a bijection. By the previous result, the map $\mathrm{Pic}(A) \to \mathrm{Pic}(A[[T]])$ is a bijection. Therefore $\mathrm{Cl}(A) \to \mathrm{Cl}(A[[T]])$ is a bijection. $\square$

As a corollary we obtain a result of Samuel [1961 a], Buchsbaum [1961] and Claborn [1965 d].

Corollary 18.23. *If A is a regular integral domain, then $\mathrm{Cl}(A)$ $\to \mathrm{Cl}(A[[T]])$ is a bijection.*

Proof. Since $A_{\mathfrak{m}}$ is a regular local ring for all maximal ideals $\mathfrak{m}$ of A, the ring $A_{\mathfrak{m}}[[T]]$ is regular and local. It is the (T)-adic completion of the local ring $A[[T]]_{\mathfrak{m}A[[T]]+(T)}$. Therefore $A[[T]]$ is a regular ring. By the Auslander and Buchsbaum theorem $A_{\mathfrak{m}}[[T]]$ is factorial. Hence the result follows from the proposition. $\square$

§ 19. Completions, Formal Power Series and Danilov's Results

In this section the two extensions completion and formal power series are studied in more detail. It has been shown in Section 6 that the induced map on the divisor class group is a monomorphism in both cases. In Section 17 on radical descent, examples have been constructed which demonstrate that these induced maps are not always bijections.

Danilov's many interesting discoveries are stated with some indication of proofs, where this is possible. As mentioned in the previous section, the problems encountered here are more geometric than algebraic. In order to go into complete detail, to completely demonstrate Danilov's results, a large amount of algebraic geometry would be needed. But this is beyond the limits of this book.

The first important result in this chapter is Scheja's theorem.

Following Danilov, say that the noetherian normal integral domain A has a *discrete divisor class group* if $i^*: \mathrm{Cl}(A) \to \mathrm{Cl}(A[[T]])$ is a bijection. Since $j_A^*: \mathrm{Cl}(A[[T]]) \to \mathrm{Cl}(A)$, as defined in Theorem 18.8, splits i^*, to say that A has a discrete divisor class group is equivalent to say that $\ker j_A^* = 0$.

Theorem 19.1 [Scheja, 1967; Storch, 1967]. *Let A be a complete local noetherian integral domain with depth $A \geqslant 3$. Suppose $\mathrm{Cl}(A)$ is a torsion group of bounded order, say n. Then $\mathrm{Cl}(A[[T]])$ is a torsion group of bounded order n.*

Proof. If $\mathrm{Cl}(A)$ is a torsion group of order n, then $n\,\mathrm{Cl}(A) = 0$ and there is an element of order n in $\mathrm{Cl}(A)$. This is equivalent to the statement that n is the least positive integer m such that $f^m A \cap g^m A$ is principal for all elements f, g in A. But $f^m A \cap g^m A$ is the kernel of the surjection $f^m A \oplus g^m A \to f^m A + g^m A$. Since A is a local ring, the ideal $f^m A \cap g^m A$ is principal if and only if the projective dimension $\mathrm{pd}_A(f^m A + g^m A) \leqslant 1$.

Let $\mathfrak{m}$ be the maximal ideal of A and let $\mathfrak{M}$ be the maximal ideal of $A[[T]]$. Then $\mathfrak{M} = \mathfrak{m} A[[T]] + T A[[T]]$. Since depth $A \geqslant 3$, the maximal ideal $\mathfrak{m}$ is not associated to any ideal of the form $f^n A + g^n A$. This is rather easy to see. Let $f^n A \cap g^n A$ be the principal ideal $h A$. Then $h = f^n r$, $h = g^n s$ and $f^n g^n = hu$ for suitable r, s and u in A. Then r, s is a regular A-sequence, the element $f^n = su$ and $g^n = ru$. Were $\mathfrak{m}$ associated to $f^n A + g^n A$, there would be a w in A such that $\mathfrak{m} w \in f^n A + g^n A$ while $w \notin f^n A + g^n A$. Since depth $A \geqslant 3$, there is a regular A-sequence of length 3 in $\mathfrak{m}$. A straightforward computation using this regular A-sequence shows that no such w can exist.

To show that $\mathrm{Cl}(A[[X]])$ is a torsion group with order n, it is sufficient to show that any ideal $\mathfrak{a} = F^n A[[T]] + G^n A[[T]]$ has projective dimension at most one. We can assume that F and G are not zero.

Let $\mathfrak{I}_j = \{H \in A[[T]] : H \mathfrak{M}^j \subseteq \mathfrak{a}\}$. Then the ascending chain of ideals $\mathfrak{I}_0 \subseteq \mathfrak{I}_1 \subseteq \cdots \subseteq \mathfrak{I}_k \subseteq \mathfrak{I}_{k+1} \subseteq \cdots$ is finite since $A[[T]]$ is noetherian. Let $\mathfrak{a}_0$ denote $\mathfrak{I}_k$ for sufficiently large k. Then $\{H \in A[[T]] : H \mathfrak{M} \subseteq \mathfrak{a}_0\} = \mathfrak{a}_0$, so $\mathfrak{M}$ is not associated to $\mathfrak{a}_0$.

Before continuing with the proof of the theorem, two additional results are needed.

Lemma 19.2 [Scheja, 1967]. *Let A be a complete local noetherian ring. Suppose $\mathfrak{P}_1, \ldots, \mathfrak{P}_r$ are nonmaximal prime ideals in $A[[T]]$. Then there is an element of the form $f + T$ with f in $\mathfrak{m}$ which is not in any of the $\mathfrak{P}_j$.*

Proof. Let $\mathfrak{Q} = \mathfrak{m} A[[T]]$, which is a prime ideal in $A[[T]]$. We can assume that each $\mathfrak{P}_j$ is distinct from $\mathfrak{Q}$ since we are searching for an element of the form $f + T$ which cannot be in $\mathfrak{Q}$ in any case. Now $\dim A[[T]]/\mathfrak{Q} = 1$, with maximal ideal $\mathfrak{M}/\mathfrak{Q}$. Hence $\mathfrak{Q}$ is not contained in any of the $\mathfrak{P}_j$. Hence $\mathfrak{P}_j \cap A \neq \mathfrak{m}$ for each $j = 1, \ldots, r$. But then $\mathfrak{m} \not\subseteq (\mathfrak{P}_1 \cap A) \cup \cdots \cup (\mathfrak{P}_r \cap A)$, so there is an f in $\mathfrak{m}$, with f not in any of the $\mathfrak{P}_j$. Suppose $f^n + T \in \mathfrak{P}_1 \cup \cdots \cup \mathfrak{P}_r$ for all positive integers n. Then there must be two integers n and m with $n < m$ and a j with $f^n + T$ and $f^m + T$ in $\mathfrak{P}_j$. But then $f^{m-n}(f^n - 1) \in \mathfrak{P}_j \cap A$ and so $f^{m-n} \in \mathfrak{P}_j \cap A$, a contradiction.   □

Lemma 19.3. *Let A be a complete local ring with maximal ideal $\mathfrak{m}$. Let f be in $\mathfrak{m}$. The composition of the ring homomorphism $A \to A[[T]]$ with $A[[T]] \to A[[T]]/(T - f)$ is a bijection.*

Proof. If $F \cdot (T - f) \in A$, then each coefficient of F is in $\bigcap_m A f^m$. Thus $F = 0$. So $A \cap (T - f) = 0$.

If $F \in A[[T]]$, then $F(f)$ is a well defined element in A since A is complete. Hence we have a surjection $A[[T]] \to A$ defined by the assignment $F \mapsto F(f)$. This factors through the epimorphism $A[[T]] \to A[[T]]/(T - f)$. ∎

Return now to the proof of the theorem. The prime ideals associated to $\mathfrak{a}_0$ are not maximal ideals. Hence there is an element of the form $f + T$ in $A[[T]]$ such that it is regular on $A[[T]]/\mathfrak{a}_0$. Let $\bar{F}$ and $\bar{G}$ denote the images of F and G in A, so that $\bar{F} = F(-f)$ and $\bar{G} = G(-f)$. Now $\bar{\mathfrak{a}} = \bar{F}^n A + \bar{G}^n A$ and is an ideal which does not have $\mathfrak{m}$ as an associated prime ideal. Hence $\mathfrak{J}_k = \{H \in A[[X]]: H \mathfrak{M}^k \subseteq \mathfrak{a} + (f + T)\} = \mathfrak{a} + (f + T)$ for all k. But $\mathfrak{J}_k \subseteq \mathfrak{J}_k$ for each k. Hence $\mathfrak{a}_0 \subseteq \mathfrak{a} + (f + T)$. Since $f + T$ is regular on $A[[T]]/\mathfrak{a}_0$, the intersection $(f + T) \cap \mathfrak{a}_0 = (f + T)\mathfrak{a}_0$. Since $\mathfrak{a} \subseteq \mathfrak{a}_0$, the modular relation gives $\mathfrak{a}_0 \cap (\mathfrak{a} + (f + T)) = \mathfrak{a} + \mathfrak{a}_0 \cap (f + T)$. Hence we get $\mathfrak{a}_0 = \mathfrak{a} + (f + T)\mathfrak{a}_0$, or $\mathfrak{a}_0 = \mathfrak{a}$. Therefore $f + T$ is regular on $A[[T]]/\mathfrak{a}$. Therefore $(f + T) \cap \mathfrak{a} = (f + T)\mathfrak{a}$.

Now $\mathrm{pd}_{A[[T]]} \mathfrak{a} = \mathrm{pd}_A(\mathfrak{a}/(f + T)\mathfrak{a})$. But $\mathfrak{a}/(f + T)\mathfrak{a} = \mathfrak{a} + (f + T)/(f + T)$, so $\mathfrak{a}/(f + T)\mathfrak{a}$ is isomorphic to $\bar{\mathfrak{a}} = \bar{F}^n A + \bar{G}^n A$. This module has projective dimension at most one as an A-module. Putting these statements together gives the desired result $\mathrm{pd}_{A[[T]]}(F^n A[[T]] + G^n A[[T]]) \leqslant 1$. ∎

Corollary 19.4. *Suppose A is a complete factorial local ring with depth $A \geqslant 3$. Then $A[[T]]$ is factorial.*

Proof. The order of the torsion group $\mathrm{Cl}(A[[T]])$ is 1 by the theorem. ∎

Corollary 19.5. *If A is a complete local factorial ring which has a discrete divisor class group, then $A[[T_1, \ldots, T_n]]$ is factorial for all integers n.*

Proof. If depth $A = 1$, then A is a principal valuation ring and so regular. The statement then follows from Proposition 18.21. If depth $A \geqslant 2$, then depth $A[[X]] \geqslant 3$. Furthermore $A[[X]]$ is a complete local factorial ring. Hence, by the Corollary 19.4, $A[[X_1, \ldots, X_n]]$ is factorial. ∎

Robbiano [1969] has extended these results to Hensel rings with depth at least three.

Having demonstrated this nice result, we give examples to show that the assumption depth $A \geqslant 3$ is necessary.

Example 19.6. (a) [Salmon, 1966]. Let k be a field, X, Y, Z, U indeterminates over k. Suppose $A = k(U)[[X, Y, Z]]/(X^2 + Y^3 + U Z^6)$.

Then A is factorial and complete, but $A[[T]]$ is not factorial. (Zinn-Justine [1967, 1968] has expanded further on this example. She shows that $A = k(U)[[X, Y, Z]]/(X^2 + Y^i + U Z^{2j})$ is factorial for all odd $(i, j) \neq (3, 3)$ while $A[[T]]$ is never factorial.)

(b) [Danilov, 1970a]. Let a be a positive integer such that $X^3 + Y^3 - a Z^3$ has only one rational zero in $\mathbf{P}^2(\mathbb{C})$. For example, the integer can be $3, 4, 5, 10, \ldots$ (see [Selmer, 1951]). Suppose p is a prime integer not dividing a. Then $A = \mathbb{Q}[x, y, z]_{(x,y,z)}$ with $p x^3 + p^2 y^3 - a z^3 = 0$ is factorial. The completion $\hat{A} = \mathbb{Q}[[x, y, z]]$ with $p x^3 + p^2 y^3 - a z^3 = 0$ is also factorial. But $\hat{A}[[T]]$ is not factorial.

To verify Salmon's example, one must show the ring is factorial but the power series ring is not factorial. There is a very nice general statement found in [Samuel, 1961a] (also [Bourbaki, 1964] and [Storch, 1967]) which shows that the power series ring in Salmon's example is not factorial.

Proposition 19.7. *Suppose A is an integral domain and x, y, z are three elements in A such that*
   (i) *x is a prime element,*
   (ii) *x, y is a regular A-sequence, and*
   (iii) *$z^{i-1} \notin y A + x A$ but $z^i \in y^j A + x^k A$.*
   (a) *If $ijk - ij - ik - jk \geqslant 0$, then $A[[T]]$ is not factorial.*
   (b) *If, moreover, A is a $\mathbb{Q}$-algebra, then $\mathrm{Cl}(A[[T]])$ is not torsion.*

Proof. There are three calculations which must be made. α) There is no power series f in $A[[T]]$ with constant term x associated to $xy - z^{i-1} T$ in $A_y[[T]]$. β) There is a power series g in $A_y[[T]]$ of the form $g = y^{-1}(x^n + b_1(T/y) + b_2(T/y)^2 + \cdots)$ for some n such that each $b_j \in A$ and $(xy - z^{i-1} T) g \in A[[T]]$. γ) If p is a prime element in A and f in $A[[T]]$ has $f(0) = p$, then f is a prime element in $A[[T]]$. These calculations can be found in [Samuel, 1961a] or [Storch, 1967].

One now shows that element $(xy - z^{i-1} T) g$ constructed in β) cannot be the product of prime elements in $A[[T]]$ and that $A[[T]]$ is not factorial. For the proof of (b) see [Storch, loc. cit.]. $\square$

Storch remarks that it is essential to assume that A is a $\mathbb{Q}$-algebra. In fact, Example 17.13 shows that the class group of $A[[T]]$ can have 2-torsion.

Two remarks are apropos here:

The function $f(i, j, k) = ijk - ik - jk$ is nonnegative for positive integer values of i, j, k unless, after some permutation, the triple (i, j, k) is one of $(2, 3, 3)$, $(2, 3, 4)$, $(2, 3, 5)$ or $(2, 2, n)$ for any n, or $(1, m, n)$ for any m and n. These rings will be discussed later.

In the section on graded rings it is seen that $k[X, Y, Z]/(X^i + Y^j + Z^k)$ is factorial for many such triples. Thus there is a wealth of examples which satisfy the properties of the proposition. The fact that the two rings in the example satisfy the properties of the proposition results from a straightforward calculation. Since Danilov, in his verification that $\mathbb{Q}[[X, Y, Z]]/(p X^3 + p^2 Y^3 - a Z^3)[[T]]$ is not factorial, uses geometrical methods, let us try to apply Samuel's result to this ring. A helpful tool in working with power series is the Preparation Theorem.

Proposition 19.8 (Weierstrass Preparation Theorem). *Let A be a complete local ring with maximal ideal* $\mathfrak{m}$. *Let F in* $A[[T]]$ *be the power series* $\sum_{i \geq 0} a_i T^i$. *Suppose there is at least one j such that* $a_j \notin \mathfrak{m}$ *and let s be the first such integer. Then there is a unit u in* $A[[T]]$ *and a polynomial* $p(T) = T^s + b_1 T^{s-1} + \cdots + b_s$ *with each* b_j *in* $\mathfrak{m}$ *such that* $F = u p$. *Moreover, the pair (u, p) is uniquely determined.*

Proof. [Bourbaki, Chap. VII, §3, no. 9, Prop. 6, 1965] ☐

Set $A = \mathbb{Q}[[X, Y, Z]]/(p X^3 + p^2 Y^3 - a Z^3)$ and denote by x, y and z the images in A of X, Y and Z respectively. Note that x is a prime element in A. To see this, it is sufficient to show that $p^2 Y^3 - a Z^3$ is a prime element in $\mathbb{Q}[[Y, Z]]$. If there is a proper factorization it can be assumed to take place in $\mathbb{Q}[[Z]][Y]$ by the Preparation theorem. A quick calculation shows that there must then be an element h in $\mathbb{Q}[[Z]]$ such that $h^3 = a/p^2$. But then the constant term of h must be a cube root of a/p^2, a contradiction.

Thus x is a prime element and then x, y is a regular A-sequence. Now $z^3 \in x^3 A + y^3 A$, whereas $z^2 \notin x A + y A$. For were z^2 in $x A + y A$, it would follow that the image of Z^2 in $\mathbb{Q}[[Y, Z]]/(p^2 Y^3 - a Z^3)$ would be in the ideal generated by the image of Y. Lifting this relation back to $\mathbb{Q}[[Y, Z]]$ would yield a relation $Z^2 - s Y = h(p^2 Y^3 - a Z^3)$ with power series s, h in $\mathbb{Q}[[Y, Z]]$. But modulo Y the relation would become $Z^2 = -h a Z^3$, which cannot occur for any h.

Thus the three conditions of Proposition 19.7 are fulfilled for the elements x, y, z in A. Now the integers i, j and k are all equal to 3. So $ijk - ij - ik - jk = 0$. Hence $A[[T]]$ is not factorial.

One consequence of the examples in 19.7 is another example.

Example 19.9. There is a factorial local ring A with maximal ideal $\mathfrak{m}$ such that the $\mathfrak{m}$-adic completion $\hat{A}$ is not factorial.

Let B be $\mathbb{Q}[x, y, z, T]$ with $p x^3 + p^2 y^3 - a z^3 = 0$. Then $A = B_{(x, y, z, T)}$ is factorial since $\mathbb{Q}[x, y, z]$ is factorial and A is a localization of the polynomial extension. Now $\hat{A} = \mathbb{Q}[[x, y, z]][[T]]$ which is not factorial.

Salmon's example works as well.

Another consequence is an example of a local noetherian ring A and a multiplicatively closed subset S such that the Picard group and the Grothendieck group of $S^{-1} A$ are not trivial [Samuel, 1961 a].

Example 19.10. There is a (complete) local noetherian ring A and a multiplicatively closed subset S such that $S^{-1}A$ is a Dedekind domain which is not factorial. Hence Pic and K_0 do not preserve epimorphisms.

For let A be one of the examples, say $A = (k(U)[[X, Y, Z]]/(X^2 + Y^3 + U Z^6))[[T]]$. Let S be the nonzero elements in $k(U)[[X, Y, Z]]/(X^2 + Y^3 + U Z^6)$. Then S meets every prime ideal of height 2 in A. Hence $S^{-1}A$ is a Dedekind domain. The set S is generated by prime elements since $S \cup \{0\}$ is a factorial ring. By Nagata's Theorem, we have the isomorphism $\mathrm{Cl}(A) \to \mathrm{Cl}(S^{-1}A)$. But $S^{-1}A$ is a Dedekind domain. Hence $\mathrm{Cl}(S^{-1}A) = \mathrm{Pic}(S^{-1}A)$. Therefore $\mathrm{Pic}(S^{-1}A) \neq 0$, whereas $\mathrm{Pic}(A) = 0$. Now $K_0(A) = \mathbb{Z}$, whereas $K_0(S^{-1}A) = \mathbb{Z} \oplus \mathrm{Pic}(S^{-1}A)$.

Among the many positive results concerning the 2-dimensional factorial rings for which the power series rings are also factorial, are the following results of Brieskorn [1968], Lipman [1969] and Scheja [1967].

The first result is a simplification of results found in [Scheja, 1967].

Proposition 19.11. *Let B be a three dimensional regular local ring with maximal ideal generated by x, y, z. Let $A = B/(ax^2 + by^3 + cz^5)$ for units a, b, c in B. Then $A[[T_1, \ldots, T_n]]$ is factorial for all n.*

Proof. [Scheja, 1967] ⬜

A very strong related result, proved first by Brieskorn [1968] for residue field $\mathbb{C}$ and later by Lipman for residue field algebraically closed with characteristic different from 2, 3 or 5, states that the above rings are essentially the only nonregular factorial rings with this property.

Before stating the result, the notion of the Henselization of a local ring is required. A local ring A is a *Hensel ring* if Hensel's lemma is satiesfied for A. This is equivalent to saying that every A-algebra which is finitely generated as an A-module is a finite product of local A-algebras (see [Raynaud, 1970]).

If A is a local ring, then a Henselization of A is a pair $(^hA, i)$, where ^hA is a local Hensel ring, $i: A \to {}^hA$ is a local homomorphism, which satisfies the universal property that any local homomorphism $j: A \to B$, where B is a local Hensel ring, factors uniquely through i. Then $(^hA, i)$ is unique up to isomorphism.

The result of Lipman can now be stated.

Proposition 19.12. *Let A be a two dimensional local ring with maximal ideal $\mathfrak{m}$ whose residue field is algebraically closed with characteristic different from 2, 3 or 5. Suppose A is not regular. Then the following conditions are equivalent.*

(i) *The completion $\hat{A}$ is factorial.*

(ii) *The completion $\hat{A}$ is normal and the Henselization $^h A$ is factorial.*

(iii) *There is a basis x, y, z of $\mathfrak{m}$ and units a and b such that $x^2 + a y^3 + b z^5 = 0$.*

(iv) *There is a basis x_h, y_h, z_h of $^h \mathfrak{m}$ (the maximal ideal of $^h A$) such that $x_h^2 + y_h^3 + z_h^5 = 0$.*

(v) *There is a three dimensional regular local ring B with maximal ideal generated by u, v, w such that $\hat{A} = B/(u^2 + v^2 + w^5)$.*

Proof. [Lipman, Theorem 25.1, 1969] □

Lipman also handles the three special cases of characteristic 2, 3, and 5. The resulting relation in condition (iii) is changed slightly. The interested reader is referred to Lipman's paper.

The results reported above more or less handle the case of the triple (2, 3, 5) which arises in Proposition 19.7. The other cases are considered by Scheja [1966]. Again simplifications of his results are mentioned.

Proposition 19.13. *Let B be a regular three dimensional local ring whose residue field is formally real. Let x, y, z be elements which generate the maximal ideal. Let a, b, c be units in B and set $A = B/(a x^i + b y^j + c z^k)$.*

(a) *If $(i, j, k) = (2, 3, 4)$ or $(2, 2, n)$, then $A[[T_1, ..., T_m]]$ is factorial for all m.*

(b) *If, in addition, the element $b - 1$ is in the maximal ideal and c has no cube root in the residue field, and if $A = B/(a x^2 + b y^3 + c z^3)$, then $A[[T_1, ..., T_m]]$ is factorial for all m.*

Proof. [Scheja, Satz 5, 6, 7, 1966] □

In the Grothendieck-Samuel Theorem, Proposition 18.15, and the theory of graded Krull domains, we saw local to global results. The next result, and its corollary, is another example of this phenomenon.

Let A be a noetherian complete local ring with maximal ideal $\mathfrak{m}$. Set $B = A[[T_1, ..., T_n]]$. Suppose A (and hence B) is normal.

Proposition 19.14 (Ramanujam-Samuel). *The surjection $Cl(B) \to Cl(B_{\mathfrak{m}B})$ is a bijection.*

Proof. The kernel of the homomorphism $Cl(B) \to Cl(B_{\mathfrak{m}B})$ is generated by the classes of those divisorial prime ideals $\mathfrak{p}$ of B such that $\mathfrak{p} \cap (B - \mathfrak{m}B) \neq \emptyset$, that is, by those $\mathfrak{p}$ not contained in $\mathfrak{m}B$. The problem is to show that such a prime ideal is principal.

Let $\mathfrak{p}$ be a divisorial prime ideal such that $\mathfrak{p} \not\subseteq \mathfrak{m}B$. Suppose $f_1, ..., f_r, g_1, ..., g_s$ are elements of B which generate $\mathfrak{p}$ and are such that $f_j \notin \mathfrak{m} B$ for each j while $g_i \in \mathfrak{m} B$ for each i. Suppose that $f_1 \neq 0$. Then the elements $f_1, \ldots, f_r, g_1 - f_1, \ldots, g_s - f_1$ generate $\mathfrak{p}$ and none of them are in $\mathfrak{m} B$. So we may assume $\mathfrak{p} = f_1 B + \cdots + f_r B$ with $f_i \notin \mathfrak{m} B$ for all i.

Let f be the product of the f_i. There is an A-automorphism u of B induced by the assignments $T_i \mapsto T_i + T_n^{t(i)}$ for $i = 1, \ldots, n-1$ such that f^u and hence each f_i^u is regular in T_n (i.e. $f^u(0, \ldots, 0, T_n) \neq 0$). By the Preparation Theorem 19.8 each f_j is the product of a unit in B times a unitary polynomial in T_n whose coefficients are in the radical of $A[[T_1, \ldots, T_{n-1}]]$. Let $\mathfrak{q} = \mathfrak{p}^u \cap A[[T_1, \ldots, T_{n-1}]][T_n]$. Then $\mathfrak{p}^u = \mathfrak{q} B$. Since B is flat as a module over $A[[T_1, \ldots, T_{n-1}]][T_n]$, the restriction $\mathfrak{q}$ is a divisorial prime ideal.

Now $\text{Cl}(A[[T_1, \ldots, T_{n-1}]][T_n]) \to \text{Cl}(A[[T_1, \ldots, T_{n-1}]](T_n))$ is a bijection by Proposition 8.9. Since $\mathfrak{q} \not\subseteq \mathfrak{m} A[[T_1, \ldots, T_{n-1}]][T_n]$, the ideal $\mathfrak{q}$ must be principal. Hence the extension $\mathfrak{p}^u$ is a principal ideal and therefore $\mathfrak{p}$ itself is principal. $\square$

Recall that a local ring A is *formally normal* (or *analytically normal*) if the $\mathfrak{m}(A)$-adic completion of A is normal. We have seen that in this case A itself is normal and that the homomorphism $\text{Cl}(A) \to \text{Cl}(\hat{A})$ is an injection.

Corollary 19.15. *Suppose A is a formally normal local ring with maximal ideal $\mathfrak{m}$. Then $\text{Cl}(A[[T_1, \ldots, T_n]]) \to \text{Cl}(A[[T_1, \ldots, T_n]]_{\mathfrak{m} A[[T_1, \ldots, T_n]]})$ is a bijection.*

Proof. Set $B = A[[T_1, \ldots, T_n]]$ and $\mathfrak{M} = \mathfrak{m} B + (T_1, \ldots, T_n)$. Then $\mathfrak{M}$ is the maximal ideal of B and the $\mathfrak{M}$-adic completion of B is $\hat{A}[[T_1, \ldots, T_n]]$. The diagram

$$
\begin{array}{ccc}
\text{Cl}(B) & \longrightarrow & \text{Cl}(\hat{B}) \\
\downarrow & & \downarrow \\
\text{Cl}(B_{\mathfrak{m} B}) & \longrightarrow & \text{Cl}(\hat{B}_{\mathfrak{m} B})
\end{array}
$$

is commutative. The completion $\hat{B}$ is normal since A is formally normal.

By Nagata's Theorem $\text{Cl}(B) \to \text{Cl}(B_{\mathfrak{m} B})$ is a surjection. Since all the homomorphisms in the diagram, except possibly this one, are injections, this map is also an injection. Hence it is a bijection. $\square$

Danilov [1968, 1970a, 1970b] has extensively studied the homomorphism $i^* : \text{Cl}(A) \to \text{Cl}(A[[T]])$. In particular, when the residue class fields of A have characteristic 0, he has given a complete description of those rings A for which this map is a bijection. The next few paragraphs represent an attempt to understand Danilov's results.

The most important tool is the retract $j^*: \mathrm{Cl}(A[[T]]) \to \mathrm{Cl}(A)$ of the injection i^*, which is shown to exist in Theorem 18.8. One would hope to be able to work with affine open subschemes of $\mathrm{Spec}\, A$ (or $\mathrm{Spec}\, A[[X]]$). But there are examples (see Section 15) of Dedekind domains which are not factorial but such that A_f is factorial for each nonunit f in A. Therefore one is necessarily forced to study Picard groups of open, but not necessarily affine, subschemes of $\mathrm{Spec}\, A$ and $\mathrm{Spec}\, A[[T]]$.

Recall again that A is said to have a discrete divisor class group if $i^*: \mathrm{Cl}(A) \to \mathrm{Cl}(A[[T]])$ is a bijection. This is equivalent to saying that j^* has a trival kernel. That A has a discrete divisor class group is reflected by faithfully flat extensions and by locally having this property, as is seen by the next result.

Proposition 19.16. (a) *Suppose* $A \to B$ *is faithfully flat, that* B *is normal and noetherian and that* B *has a discrete divisor class group. Then* A *has a discrete divisor class group.*

(b) *Suppose* $A_{\mathfrak{m}}$ *has a discrete divisor class group for each maximal ideal* $\mathfrak{m}$ *of* A. *Then* A *has a discrete divisor class group.*

The proof depends on a lemma. Suppose A is a normal noetherian ring. Identify the affine scheme $\mathrm{Spec}\, A$ with the closed subscheme $V(T)$ of $\mathrm{Spec}\, A[[T]]$. Each open subscheme U of $\mathrm{Spec}\, A$ is then the intersection of an open subscheme V of $\mathrm{Spec}\, A[[T]]$ with $\mathrm{Spec}\, A$. Let $f_U: U \to \mathrm{Spec}\, A[[T]]$ be the composition of the morphisms $U \to \mathrm{Spec}\, A$ and $\mathrm{Spec}\, A \to \mathrm{Spec}\, A[[T]]$. Recall also that $U_{\mathfrak{a}} = \{x \in \mathrm{Spec}\, A[[T]]: \mathfrak{a}_x$ is invertible$\}$.

Lemma 19.17. *Suppose* $\mathfrak{a}$ *is a divisorial ideal of* $A[[T]]$ *such that* $j^*(\mathrm{cl}(\mathfrak{a})) = 0$. *If* $U \subseteq U_{\mathfrak{a}}$, *then* $f_U^*(\tilde{\mathfrak{a}}) \cong \mathcal{O}_U$. *If* $A \xrightarrow{g} B$ *is flat and* B *has a discrete divisor class group, then* $U_{\mathfrak{a}}$ *contains* $(\mathrm{Spec}(g)(\mathrm{Spec}\, B))$.

Proof. Suppose $U' = U_{\mathfrak{a}} \cap \mathrm{Spec}\, A$. If $x \in \mathrm{Spec}\, A - U'$, then $\dim A_x \geq 2$. Furthermore $U \subseteq U'$, and the map f_U is the composition of $U \xrightarrow{i} U'$ with $f_{U'}$. Also, the restriction $i^*(\mathcal{O}_{U'}) = \mathcal{O}_U$. Hence we may assume that $\dim A_x \geq 2$ for x in $\mathrm{Spec}\, A - U$. Now $\mathrm{Pic}(U) \to \mathrm{Cl}(A)$ is an injection by Corollary 18.5. The element $j^*(\mathrm{cl}(\mathfrak{a}))$ is the image in $\mathrm{Cl}(A)$ of the element defined by $f_U^*(\tilde{\mathfrak{a}})$ in $\mathrm{Pic}(U)$. Since this element is trivial, the sheaf $f_U^*(\tilde{\mathfrak{a}}) \cong \mathcal{O}_U$.

Now suppose $A \xrightarrow{g} B$ is flat and that B is normal, noetherian and has a discrete divisor class group. The diagram

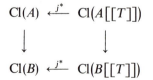

commutes. Now $j^*(\mathrm{cl}(\mathfrak{a}\,B[[T]]))=0$ and hence $\mathrm{cl}(\mathfrak{a}\,B[[T]])=0$. Therefore $\mathfrak{a}\,B[[T]]$ is a principal ideal.

Let $\mathfrak{P}$ be a prime ideal in B and denote by $\mathfrak{p}$ the prime ideal $\mathfrak{P}\cap B$, which is the image under $\mathrm{Spec}(g)$ of $\mathfrak{P}$. The inverse image of $\mathfrak{p}$ in $A[[T]]$ is the prime ideal $\mathfrak{p}A[[T]]+(T)$, which we denote by $\mathfrak{q}$. Let $\mathfrak{Q}$ denote the prime ideal $\mathfrak{P}B[[T]]+(T)$. Then $\mathfrak{Q}\cap A[[T]]=\mathfrak{q}$, and the homomorphism $A[[T]]_{\mathfrak{q}}\to B[[T]]_{\mathfrak{Q}}$ is faithfully flat. Since $\mathfrak{a}B[[T]]_{\mathfrak{Q}}$ is a principal ideal, the ideal $\mathfrak{a}A[[T]]_{\mathfrak{q}}$ is invertible. Hence $\mathfrak{q}\in U_{\mathfrak{a}}$ and so $\mathrm{Spec}(g)(\mathfrak{P})\in U_{\mathfrak{a}}$.

Proposition 19.16 can now be demonstrated.

Proof of Proposition 19.16. (a) Suppose $A\to B$ is faithfully flat and that B has a discrete divisor class group. Suppose $j^*(\mathrm{cl}(\mathfrak{a}))=0$ for the divisorial ideal $\mathfrak{a}$ of $A[[T]]$. Then, as in the proof above, the extension $\mathfrak{a}B[[T]]$ is an invertible ideal. Hence $\mathfrak{a}$ is invertible, since $A[[T]]\to B[[T]]$ is faithfully flat. But then $\mathfrak{a}\otimes_{A[[T]]}A\cong\mathfrak{a}/T\mathfrak{a}$ which is isomorphic to A, by the lemma. Hence $\mathfrak{a}$ is itself free and so a principal ideal.

(b) This proof follows closely the proof of Theorem 3.1 of [Samuel, 1961]. Suppose $\mathfrak{a}$ is a divisorial ideal such that $j^*(\mathrm{cl}(\mathfrak{a}))=0$. Let $\mathfrak{M}$ be a maximal ideal of $A[[T]]$. Then $\mathfrak{M}=\mathfrak{m}A[[T]]+(T)$ for a maximal ideal of A. The local ring $A_{\mathfrak{m}}[[T]]$ is the (X)-adic completion of the Zariski ring $A[[T]]_{\mathfrak{M}}$. Therefore $A[[T]]_{\mathfrak{M}}\to A_{\mathfrak{m}}[[T]]$ is faithfully flat. Since $j^*(\mathrm{cl}(\mathfrak{a}))=0$, it follows that $j^*(\mathrm{cl}(\mathfrak{a}A_{\mathfrak{m}}[[T]]))=0$. But $A_{\mathfrak{m}}[[T]]$ has a discrete divisor class group. Therefore the divisorial ideal $\mathfrak{a}A_{\mathfrak{m}}[[T]]$ is principal. Hence the extension of $\mathfrak{a}$ to $A[[T]]_{\mathfrak{M}}$ is principal. But then $\mathfrak{a}$ is a projective finitely generated ideal and so invertible. As $j^*(\mathrm{cl}(\mathfrak{a}))=\mathrm{cl}(\mathfrak{a}/T\mathfrak{a})$, in this case, and as $\mathfrak{a}/T\mathfrak{a}$ is free, it follows that $\mathfrak{a}$ is principal. □

Part (b) of the proposition is a generalization of Samuel's Theorem 3.1 [1961 a].

It is not clear that the property of having a discrete divisor class group is preserved by ring of quotient formation.

Problem 19.18. Suppose A has a discrete divisor class group and S is a multiplicatively closed subset of A. Does $S^{-1}A$ have a discrete divisor class group?

For excellent rings with residue class fields of characteristic 0, the answer to this problem is affirmative.

(The ring A is *excellent* if

(a) it is noetherian,

(b) the polynomial rings $A[T_1,\ldots,T_n]$ are catenary for each finite set of indeterminates $T_1,\ldots,T_n$ (i.e., A is universally catenary),

(c) for each homomorphism $\alpha: A \to K$ to a field K, the ring $(A_{\ker \alpha})^{\wedge} \otimes_A K$ is regular and

(d) for each finitely generated A-algebra B, the set $\{\mathfrak{p}: \mathfrak{p} \in \operatorname{Spec} B$ and $B_{\mathfrak{p}}$ is regular$\}$ is open in $\operatorname{Spec} B$.

The class of excellent rings is stable under localization and base extension to finitely generated algebras. Dedekind domains of characteristic zero are excellent. But there are principal valuation rings of characteristic p (for any $p \neq 0$) which are not excellent. Complete semilocal noetherian rings are excellent. Excellent local rings are formally (i.e., analytically) normal.)

Proposition 19.19. *Suppose A is an excellent normal noetherian domain whose residue fields have characteristic 0. Then A has a discrete divisor class group if and only if $S^{-1}A$ has a discrete divisor class group for all multiplicatively closed subsets S of A.*

Proof. [Danilov, 1970a, 1970b] $\quad \Box$

The proof that $S^{-1}A$ has a discrete divisor class group uses the fact that $\operatorname{Spec} A$ (and hence also $\operatorname{Spec} A[[T]]$) admits a resolution of singularities [Hironaka, 1967]. Of course the other direction follows from Proposition 19.17b.

As part of the proof one is led to consider a projective birational morphism $f: Y \to \operatorname{Spec} A$ where Y is a regular integral scheme (a *resolving morphism*).

Proposition 19.20. *The ring A has a discrete divisor class group if and only if $\mathrm{H}^1(Y, \mathcal{O}_Y) = 0$.*

Proof. [Danilov, 1970b, pages 375—376] $\quad \Box$

Corollary 19.21. *Suppose A is a normal resolvable ring of Krull dimension 2. Suppose B is an A-algebra such that $\operatorname{Spec} B \to \operatorname{Spec} A$ is regular. If A has a discrete divisor class group, then so does B.* $\quad \Box$

Corollary 19.22. *If A is a normal resolvable local ring, then its completion $\hat{A}$ is normal. Furthermore A has a discrete divisor class group if and only if $\hat{A}$ has a discrete divisor class group.* $\quad \Box$

Recall that an A-algebra B is *regular* over A if B is flat as an A-module and for any homomorphism to a field $A \to K$, the ring $B \otimes_A K$ is regular.

In order to facilitate the formulation of the remaining results, call the ring A *allowable* (resp; *locally allowable*) if A is either a factor ring of a Cohen-Macaulay ring, a formally normal ring or an excellent ring (resp; $A_{\mathfrak{p}}$ is *allowable* for each $\mathfrak{p} \in \operatorname{Spec} A$).

The next lemma is crucial for the remainder of Danilov's results.

Lemma 19.23. *Suppose A is a local normal allowable noetherian domain with* $\dim A \geqslant 3$. *Suppose* $U = \operatorname{Spec} A - \{\mathfrak{m}\}$ *and that $\mathfrak{a}$ is a divisorial ideal. Then* $\mathrm{H}^0(U, \tilde{\mathfrak{a}})$ *and* $\mathrm{H}^1(U, \tilde{\mathfrak{a}})$ *are A-modules of finite type.*

Proof. Since $\mathfrak{a}$ is divisorial and since $\dim A \geqslant 3$, the map $\mathfrak{a} \to \mathrm{H}^0(U, \tilde{\mathfrak{a}})$ is a bijection. (In fact, $\mathrm{H}^0(U, \tilde{\mathfrak{a}}) = \bigcap_{x \in U} \mathfrak{a}_x$, but this intersection is just $\mathfrak{a}$ since $\mathfrak{a}$ is divisorial.)

Let B be the $\mathfrak{m}$-adic completion of A. Denote by V the punctured neighborhood of $\{\mathfrak{m} B\}$. Since $A \to B$ is faithfully flat, there is an isomorphism $\mathrm{H}^1(V, \tilde{\mathfrak{a}}_B) = \mathrm{H}^1(U, \tilde{\mathfrak{a}}) \otimes_A B$. It is sufficient to prove that $\mathrm{H}^1(V, \tilde{\mathfrak{a}}_B)$ is finitely generated since faithfully flat morphisms descend finite generation.

Since B is a complete local ring, it is sufficient to show that $\operatorname{depth}(\mathfrak{a} B)_{\mathfrak{p}} \geqslant 2$ for any prime ideal of B such that $\dim B/\mathfrak{p} = 1$ [Grothendieck, 1963, exp. VIII, Theorem II-3].

In case A is formally normal, which includes the case that A is excellent, the completion B is normal. But B is catenary, by the Cohen structure theorem, and it is an integral domain. Hence $\dim B/\mathfrak{p} = 1$ implies that $\operatorname{ht} \mathfrak{p} \geqslant 2$. Since $\mathfrak{a} B$ is divisorial, we get $\operatorname{depth}(\mathfrak{a} B)_{\mathfrak{p}} \geqslant 2$.

If A is a factor ring of a Cohen-Macaulay ring, then so is the completion B. Hence B is catenary and equidimensional. Then $\dim B/\mathfrak{p} = 1$ implies $\dim B_{\mathfrak{p}} \geqslant 2$. Since the formal fibres of A satisfy (S_2), we get $\operatorname{depth}(\mathfrak{a} B)_{\mathfrak{p}} \geqslant \inf(2, \dim B_{\mathfrak{p}}) \geqslant 2$.

Hence, in either case, the conditions of [Grothendieck, loc. cit.] are satisfied. So $\mathrm{H}^1(U, \tilde{\mathfrak{a}})$ is finitely generated. $\quad\square$

One main application of this lemma is to prove the next result.

Lemma 19.24. *Suppose A is a local, normal, allowable integral domain with* $\operatorname{depth} A \geqslant 3$. *If $A_{\mathfrak{p}}$ has a discrete divisor class group for each non-maximal prime ideal $\mathfrak{p}$, then A itself has a discrete divisor class group.*

Proof. Let $\mathfrak{m}$ denote the maximal ideal of A. The maximal ideal $\mathfrak{M}$ of $A[[T]]$ is then $\mathfrak{m} A[[T]] + (T)$. Let X denote the spectrum $\operatorname{Spec} A[[T]]$ and U the punctured neighborhood of $\mathfrak{M}$. So $U = \operatorname{Spec} A[[T]] - \{\mathfrak{M}\}$. Suppose $\mathfrak{a}$ is a divisorial ideal in $A[[T]]$ for which $j^*(\operatorname{cl}(\mathfrak{a})) = 0$. It must be shown that $\mathfrak{a}$ is principal. Let $\mathscr{F}$ denote the sheaf of Ideals defined by $\tilde{\mathfrak{a}}$, so $\mathscr{F} = \tilde{\mathfrak{a}}$. The element T is regular on $\mathscr{F}$, so there is the exact sequence $0 \longrightarrow \mathscr{F} \xrightarrow{T} \mathscr{F} \longrightarrow \mathscr{F}/T\mathscr{F} \longrightarrow 0$. Hence the restriction to U gives an exact sequence $0 \longrightarrow \mathscr{F}_U \xrightarrow{T} \mathscr{F}_U \longrightarrow (\mathscr{F}/T\mathscr{F})_U \longrightarrow 0$. The cohomology long exact sequence is

$$0 \longrightarrow \mathrm{H}^0(U, \mathscr{F}) \xrightarrow{T} \mathrm{H}^0(U, \mathscr{F}) \longrightarrow \mathrm{H}^0(U, \mathscr{F}/T\mathscr{F}) \longrightarrow$$
$$\longrightarrow \mathrm{H}^1(U, \mathscr{F}) \xrightarrow{T} \mathrm{H}^1(U, \mathscr{F}) \longrightarrow \mathrm{H}^1(U, \mathscr{F}/T\mathscr{F}) \longrightarrow \cdots.$$

For each $\mathfrak{p}$ in $\operatorname{Spec} A - \{\mathfrak{m}\}$, the local ring $A_\mathfrak{p}$ has a discrete divisor class group. Since $j^*(\operatorname{cl}(\mathfrak{a})) = 0$, the open set $U_\mathfrak{a}$ of points in X at which $\mathfrak{a}$ is invertible contains $\{\mathfrak{p} A[[T]] + (T) : \mathfrak{p} \in \operatorname{Spec} A - \{\mathfrak{m}\}\}$. Hence $(\mathscr{F}/T\mathscr{F})_U \cong (\mathcal{O}_X/T\mathcal{O}_X)_U$ by Lemma 19.17. Since $\operatorname{depth} \mathcal{O}_X/T\mathcal{O}_X = \operatorname{depth} A$ and $\operatorname{depth} A \geqslant 3$ by hypothesis, we get $\operatorname{depth} \mathcal{O}_X/T\mathcal{O}_X \geqslant 3$. Hence $\mathrm{H}^0(U, \mathscr{F}/T\mathscr{F}) = A$ and $\mathrm{H}^1(U, \mathscr{F}/T\mathscr{F}) = 0$.

By Lemma 19.24, the group $\mathrm{H}^1(U, \mathscr{F})$ is a finitely generated $A[[T]]$-module. By Nakayama's Lemma $\mathrm{H}^1(U, \mathscr{F}) = 0$. Hence $0 \longrightarrow \mathrm{H}^0(U, \mathscr{F}) \overset{T}{\longrightarrow} \mathrm{H}^0(U, \mathscr{F}) \longrightarrow A \longrightarrow 0$ is exact. But $\mathrm{H}^0(U, \mathscr{F}) = \mathfrak{a}$ since $\operatorname{depth} \mathfrak{a} \geqslant 2$. Hence $0 \longrightarrow \mathfrak{a} \overset{T}{\longrightarrow} \mathfrak{a} \longrightarrow A \longrightarrow 0$ is exact. Since T is a radical and regular element in $A[[T]]$, the ideal $\mathfrak{a}$ must be free and hence principal. $\square$

This climbing procedure is now used to prove the important theorem.

Theorem 19.25. *Suppose A is a noetherian normal integral domain which is locally allowable. If $A_\mathfrak{p}$ has a discrete divisor class group for all those prime ideals $\mathfrak{p}$ with $\operatorname{depth} A_\mathfrak{p} = 2$, then A has a discrete divisor class group.*

Proof. According to Proposition 19.16(b), it is sufficient to show that each local ring $A_\mathfrak{p}$ has a discrete divisor class group. Suppose there is a prime ideal $\mathfrak{p}$ such that $A_\mathfrak{p}$ does not have a discrete divisor class group. Since A is normal, the ideal cannot have height 0 or 1. Then by hypothesis $\operatorname{depth} A_\mathfrak{p} \geqslant 3$. Let $\mathfrak{p}$ be minimal in the set $\{\mathfrak{q} : \mathfrak{q} \in \operatorname{Spec} A$ and $A_\mathfrak{q}$ does not have a discrete divisor class group$\}$.

Then $\operatorname{depth} A_\mathfrak{p} \geqslant 3$. The local ring $A_\mathfrak{p}$ is allowable. If $\mathfrak{q} \in \operatorname{Spec} A_\mathfrak{p} - \{\mathfrak{p} A_\mathfrak{p}\}$, Then $A_\mathfrak{q}$ has a discrete divisor class group. According to the previous lemma $A_\mathfrak{p}$ has a discrete divisor class group. This is a contradiction and the result is established. $\square$

Of course, we have seen local noetherian Cohen-Macaulay rings of dimension 2 which do not have a discrete divisor class group. Thus that assumption cannot be weakened or omitted in the theorem. The assumption that the ring is locally allowable is not very strong.

Another of Samuel's results, Theorem 3.2 [1961a], is a corollary of this theorem.

Corollary 19.26. *Suppose A is locally Cohen-Macaulay and that $A_\mathfrak{p}[[T]]$ is factorial if $\operatorname{ht} \mathfrak{p} = 2$. Then $\operatorname{Cl}(A) \to \operatorname{Cl}(A[[T]])$ is a bijection.*

Proof. The ring A is certainly locally allowable. Furthermore $\operatorname{depth} A_\mathfrak{p} = 2$ if and only if $\operatorname{ht} \mathfrak{p} = 2$. In this case $A_\mathfrak{p}$ has a discrete divisor class group. Therefore so does A. $\square$

Corollary 19.27. *Suppose A satisfies conditions (S_3) and (R_2) and is locally allowable. Then A has a discrete divisor class group.*

Proof. Since A satisfies (S_3), the local ring A_p has depth $A_p = 2$ if and only if ht $p = 2$. If ht $p = 2$, the localization A_p is regular by condition (R_2). But regular local rings have discrete divisor class groups. Hence all the conditions of the theorem are fulfilled. So A has a discrete divisor class group. ☐

Before considering the next results we define the *strict Henselization* ^{sh}A of a local ring A with maximal ideal $\mathfrak{m}$ and residue field k. [EGA, IV, 18, no. 8] The ring ^{sh}A is defined, up to isomorphisn, by the properties
   (a) ^{sh}A is a Hensel local A-algebra with maximal ideal $\mathfrak{m}\,^{sh}A$.
   (b) $^{sh}A/\mathfrak{m}\,^{sh}A$ is the separable closure of k.
   (c) If B is a local A-algebra which is Henselian and with a separably closed residue class field, there is a unique local A-algebra homomorphism $^{sh}A \rightarrow B$.
   The strict Henselization of A satisfies the properties listed below:
   (i) The structure morphism $A \rightarrow\ ^{sh}A$ is faithfully flat.
   (ii) The induced morphism $\operatorname{Spec}\,^{sh}A \rightarrow \operatorname{Spec} A$ has geometrically regular fibres.
   (iii) The strict Henselization ^{sh}A is noetherian (resp. normal) if and only if A is noetherian (resp. normal).
   Danilov [1970a] calls a ring A *geometrically factorial* if its strict Henselization ^{sh}A is factorial. There he proves that a complete normal equicharacteristic local noetherian ring with isolated and resolvable singularity and perfect residue field transfers geometric factoriality to the power series ring. As concerns two dimensional normal local rings with resolvable singularity, Danilov [1970b] can relate the discrete divisor group property to the finiteness of the class group of the strict Henselization.

Proposition 19.28. *Suppose A is a noetherian local normal ring with perfect residue class field. Suppose further that A has a resolvable singularity. Then A has a discrete divisor class group if and only if $Cl(^{sh}A)$ is finite.*

Proof. [Danilov, 1970b, Prop. 8] ☐

In proving this result Danilov uses the two corollaries to Proposition 19.20.
For local rings of dimension at least three Danilov considers the sheaf of Witt vectors W_∞ [Mumford, 1966, lecture 26].

Proposition 19.29. *Suppose A is a local normal allowable ring with $\dim A \geqslant 3$.*
   (a) If $H^1(\operatorname{Spec} A - \{\mathfrak{m}\}, W_\infty) \neq 0$, then A does not have a discrete divisor class group.

(b) *If* $H^1(\operatorname{Spec} A - \{\mathfrak{m}\}, W_\infty) = 0$ *and* $A_\mathfrak{p}$ *has a discrete divisor class group for* $\mathfrak{p} \neq \mathfrak{m}$, *then* A *has a discrete divisor class group.*

Proof. [Danilov, 1970 b, Prop. 5]   □

Corollary 19.30. *Suppose* A *is a local normal allowable ring with* $\dim A \geqslant 3$ *and* $\operatorname{depth} A = 2$. *If the characteristic of the residue class field is zero, then* A *does not have a discrete divisor class group.*   □

Example 19.31. Suppose X is an abelian variety over $\mathbb{C}$. Let $\mathscr{L}$ be a very ample invertible $\mathscr{O}_X$-Module. Then set $A = \coprod_{n \geqslant 0} H^0(X, \mathscr{L}^n)$. Then the local ring at the vertex of the cone $A_\mathfrak{m}$ has depth equal to two and dimension at least three. It clearly satisfies the other properties and so has not a discrete divisor class group.

The assumption that the characteristic of the residue class field be zero is crucial, as is demonstrated by Danilov. The example is similar to the above. Let X be a projective nonsingular variety defined over a field of k characteristic p, with $p \geqslant 5$ such that $\dim_k H^1(X, \mathscr{O}_X) = 1$ and the Bockstein operation $\beta_0 : H^1(X, \mathscr{O}_X) \to H^2(X, \mathscr{O}_X)$ is nonzero. Let $\mathscr{L}$ be a very ample invertible $\mathscr{O}_X$-Module such that $H^1(X, \mathscr{L}^n) = 0$ for $n > 1$. Let $\coprod_{n \geqslant 0} H^0(X, \mathscr{L}^n)$. Then B is normal (by Serre's criterion). Let A be the local ring at the vertex of the cone, so $A = B_\mathfrak{m}$. Then $\operatorname{Cl}(A) \to \operatorname{Cl}(A[[T]])$ is a bijection by results in [Danilov, 1968].

Proposition 19.32. *Suppose* A *is a* $\mathbb{Q}$-*algebra which is an excellent normal ring. Then* A *has a discrete divisor class group if and only if*
(a) *the ring* A *has the Serre property* (S_3) *and,*
(b) *if* $\operatorname{ht} \mathfrak{p} = 2$, *then* $\operatorname{Cl}(^{sh}A_\mathfrak{p})$ *is finite.*

Proof. Suppose A satisfies conditions (a) and (b). According to [Hironaka, 1964], the rings $A_\mathfrak{p}$ are resolvable for $\operatorname{ht} \mathfrak{p} = 2$. Since $\operatorname{depth} A_\mathfrak{p} \geqslant \inf(\operatorname{ht} \mathfrak{p}, 3)$, the local ring $A_\mathfrak{p}$ has $\operatorname{depth} A_\mathfrak{p} = 2$ if $\operatorname{ht} \mathfrak{p} = 2$. By Proposition 19.28, the local ring $A_\mathfrak{p}$ has a discrete divisor class group. Thus by Theorem 19.25, the ring A has a discrete divisor class group.

If, on the other hand, the ring A has a discrete divisor class group, then by the localization result, Proposition 19.19, each local ring $A_\mathfrak{p}$ has also a discrete divisor class group. If $\operatorname{ht} \mathfrak{p} \geqslant 3$, then $\operatorname{depth} A_\mathfrak{p} \geqslant 3$ by Corollary 19.30. Therefore A has property (S_3). If $\operatorname{ht} \mathfrak{p} = 2$, then $\operatorname{Cl}(^{sh}A_\mathfrak{p})$ is finite by Proposition 19.28.   □

Finally for a nice class of rings A, the property of having a discrete divisor class group ascends to regular A-algebras.

Theorem 19.33. *Suppose A is a normal locally allowable ring such that*

(a) *for each prime ideal $\mathfrak{p}$ in $\operatorname{Spec} A$, the local ring $A_{\mathfrak{p}}$ has a discrete divisor class group, and*

(b) *it is resolvable if $\dim A_{\mathfrak{p}} = 2$.*

Suppose B is a regular A-algebra which is locally allowable. Then B has properties (a) *and* (b). *In particular B has a discrete divisor class group.*

Proof (according to [Danilov, 1970b, Theorem 3]). Normality ascends along regular morphisms [Matsumura, 1971]. For the remainder we may assume B is local with maximal ideal $\mathfrak{n}$, that A is local with maximal ideal $\mathfrak{m}$ which is $\mathfrak{n} \cap A$ and then that $A \to B$ is faithfully flat.

By [Danilov, 1970a] B is resolvable if $\dim B = 2$. Thus property (b) is satisfied.

It remains to show that B has a discrete divisor class group. If $\dim A = 2$, then this follows from Corollary 19.21. (If $\dim A = 1$, then A is regular and $B/\mathfrak{m}B$ is regular, so B itself is regular and hence has a discrete divisor class group.)

Thus we may suppose that $\dim A \geqslant 3$. We know that $\operatorname{depth} B = \operatorname{depth} A + \operatorname{depth} B/\mathfrak{m}B$. But $B/\mathfrak{m}B$ is a regular local ring. Hence $\operatorname{depth} B/\mathfrak{m}B = \dim B/\mathfrak{m}B$. Since $\operatorname{depth} A \geqslant 2$ by Serre's criterion, we have $\operatorname{depth} B \geqslant 2 + \dim B/\mathfrak{m}B$. Therefore, if $\dim B/\mathfrak{m}B \geqslant 1$, then $\operatorname{depth} B \geqslant 3$. By Theorem 19.25, the ring B has a discrete divisor class group. In case $\dim B/\mathfrak{m}B = 0$, the local ring $B/\mathfrak{m}B$ is a separable field extension of $A/\mathfrak{m}$ since $B/\mathfrak{m}B$ is a regular $A/\mathfrak{m}$-algebra. This case is handled by another of Danilov's results which follows below.   □

Proposition 19.34. *Let $A \to B$ be a faithfully flat morphism of local rings. Let $\mathfrak{m}$ be the maximal ideal of A. Suppose*

(a) $\dim A \geqslant 3$,

(b) $B/\mathfrak{m}B$ *is a separable extension of $A/\mathfrak{m}$ and,*

(c) *For all $\mathfrak{q}$ in $\operatorname{Spec} B - \{\mathfrak{m}B\}$, the local rings $B_{\mathfrak{q}}$ have discrete divisor class groups.*

Then B has a discrete divisor class group if and only if A has.

Proof. [Danilov, 1970b, Prop. 6]   □

The applications of Theorem 19.33 are manifold. Danilov refers to several of them. The ring A is assumed to satisfy the conditions of the theorem.

Corollary 19.35. *The polynomial extension $A[X_1, \dots, X_n]$ has a discrete divisor class group.*   □

Corollary 19.36. *If A is a factor of a local Cohen-Macaulay ring, then the Henselization ${}^{h}A$ and the strict Henselization ${}^{sh}A$ have discrete divisor class groups.* $\square$

Corollary 19.37. *If A is excellent then any ideal-adic completion of A has a discrete divisor class group.* $\square$

Corollary 19.38. *Suppose A is excellent. Then*

$$\mathrm{Cl}(A) \to \mathrm{Cl}\big(A[[T_1, \ldots, T_n]]\big)$$

is a bijection for any n. $\square$

The final result of Danilov's is this last theorem.

Theorem 19.39. *Suppose A is a factor ring of a regular ring and that $A_{\mathfrak{p}}$ is regular for all prime ideals $\mathfrak{p}$ with $\mathrm{ht}\,\mathfrak{p} \leqslant 2$, and, as usual, A is normal and has a discrete divisor class group. Then $A[[T]]$ has the same properties. In particular, $\mathrm{Cl}(A) \to \mathrm{Cl}\big(A[[T_1, \ldots, T_n]]\big)$ is a bijection.*

Proof. [Danilov, 1970b, Prop. 9] $\square$

We have not stated many results from two earlier papers of Danilov. In one of them [1968] he is able to compute the class group of the completion and then the polynomial extension in terms of low dimensional cohomology groups. There is an intimate relation between these calculations and the calculations which Samuel [1964d] makes using radical descent.

Danilov has solved many of the problems in this area. But there are many interesting problems which remain.

Remarks. Wenche Melchiors [Special, Aarhus Univ. 1972] has demonstrated the existence of the splitting $j^*: \mathrm{Cl}(A[[T]]) \to \mathrm{Cl}(A)$ by purely ring theoretical arguments. Suppose $\mathfrak{a}$ is a divisorial ideal in $A[[T]]$. Let v be the (T)-adic valuation of $A[[T]]$. Define the divisorial ideal $\mathfrak{b} = T^{-v(\mathfrak{a})}\mathfrak{a}$. Let $j: A[[T]] \to A$ be given by sending $T \mapsto 0$ and extending to $A[[T]]$. Then $j(\mathfrak{b})$ is a nonzero ideal in A. Now define $j^*(\mathrm{cl}(\mathfrak{a}))$ to be $\mathrm{cl}(A : (A : j(\mathfrak{b})))$. Melchior shows that this assignment induces a well defined homomorphism, that it splits i^* and that it agrees with Danilov's j^*.

Grothendieck [Le groupe de Brauer II, Sém. Bourbaki 1965/66, no. 27 and Dix Exposés sur la Cohomologie des Schémas. Amsterdam and Paris: North-Holland and Masson et Cie 1968] mentions that Mumford [IHES Publ. Math. No. 9, 1960] gives an example of a two dimensional factorial local ring A defined over $\mathbb{C}$ whose (strict) Henselization has a nonzero torsion free class group. This is an example of a ring which does not have a discrete divisor class group (Proposition 19.28).

Grothendieck [SGA, 1962] also considers the concept "geometrically factorial" and in Exposé XIII entitled *Problemes et Conjectures* discusses some of the problems which remain in this area.

Appendix I. Terminology and Notation

Standard terminology and notation, which is used for example in [Bourbaki, 1961—1965], is employed throughout. The following short list of definitions and notations is recalled for the benefit of the reader.

Cohen-Macaulay modules: A module M of finite type over the noetherian local ring A is Cohen-Macaulay if depth $M = \dim M$. The ring itself is Cohen-Macaulay if depth $A = \dim A$.

Depth$_J$ M: If J is an ideal in A and M is an A-module of finite type, then depth$_J$ M is the length of a longest regular M-sequence contained in J. If J is the maximal ideal in the local ring A, then depth $M = $ depth$_J$ M.

Gorenstein ring: The local ring A is Gorenstein if the injective dimension, id$_A$ A, of A as a module is finite [Bass, 1963].

Normal ring: A domain is normal if it is integrally closed and conversely.

Principal valuation ring: A ring is a principal valuation ring if it is a discrete rank one valuation ring or, equivalently, a local noetherian domain whose maximal ideal is generated by at most one element.

Regular (local) ring: A noetherian ring A is regular if each localization at the maximal ideals is a regular local ring. The local noetherian ring A with maximal ideal m and residue field k is regular if $\dim A = \dim_k m/m^2$. By the Hilbert-Serre syzygy theorem, this is the same as to say that the projective dimension pd$_A k$ is finite.

Regular M-sequence: If M is an A-module of finite type and if $x_1, \ldots, x_n$ is a set of elements in A such that x_1 is not a zero divisor on M (i.e., x_1 is regular on M or M-regular) and if x_i is regular on $M/x_1 M + \cdots + x_{i-1} M$ for $1 < i \leqslant n$, then $x_1, \ldots, x_n$ is a regular M-sequence of length n.

A_f: A is commutative ring with element f. S is the multiplicatively closed subset $\{1, f, f^2, \ldots\}$. Then $S^{-1} A$ is denoted by A_f (or $A[f^{-1}]$).

A^*: This denotes the group of units of the ring A. $\mathfrak{a}^{**}$: $\mathfrak{a}^{**} = A : (A : \mathfrak{a})$.

$\mathbf{A}^n(k)$: n dimensional affine space over k.

$\mathbb{C}$: field of complex numbers.

$\mathbb{N}$: set of natural numbers (sometimes with 0, other times without).

$\mathbf{P}^n(k)$: n dimensional projective space over k.

$\mathbb{Q}$: field of rational numbers.

$\mathbb{R}$: field of real numbers.

(R_n): The ring A satisfies property (R_n) if $A_{\mathfrak{p}}$ is a regular local ring for all $\mathfrak{p}$ in Spec A with ht $\mathfrak{p} \leqslant n$.

(S_n): The ring A satisfies the Serre (S_n) property if depth $A_{\mathfrak{p}} \geqslant \inf(n, \text{ht } \mathfrak{p})$ for all $\mathfrak{p}$ in Spec A.

$\coprod$: This denotes the coproduct (or direct sum).

$X^{(p)}$: For the locally noetherian scheme X, the subset $X^{(p)} = \{x \in X : \dim \mathcal{O}_{X,x} = p\}$. If $X = $ Spec A, then $X^{(p)} = \{\mathfrak{p} \in \text{Spec } A : \text{ht } \mathfrak{p} = p\}$.

$\mathbb{Z}$: The ring of integers.

Appendix II. List of Results

In this appendix we list the numbered statements which appear in the text, following the example of a similar most useful list in [Kaplansky, 1970]. Of necessity, many statements are abbreviated, hopefully not beyond recognition.

Chapter I. Introduction to Krull Domains

§1. The Definition of a Krull Ring.
1.1 A Krull domain is integrally closed.
1.2 $A \cap k$ is a Krull domain.
1.3 The integral closure of a Krull domain in a finite extension is a Krull domain.
1.4 $\bigcap A_i$ is a Krull domain.
1.5 Subintersections are Krull domains.
1.6 $A[\{T_i\}]$ is a Krull domain.
1.7 $A[[T]]$ is a Krull domain.
1.8 $S^{-1}A$ is a Krull domain.
1.9 Minimal nonzero prime ideal characterization of a Krull domain.
1.10 Example of a Krull domain with a minimal prime ideal which is not finitely generated.

§2. Lattices.
2.1 Equivalent characterizations of a lattice.
2.2 Closure properties of lattices.
2.3 $N \subseteq N'$ and $M' \subseteq M$ implies $N:M \subseteq N':M'$.
2.4 $A:M = A:(A:(A:M))$ for an A-lattice M.
2.5 $A:(A:M) = \bigcap_{\alpha \in A:M} \alpha^{-1}(A) \cap F$.
2.6 If M is divisorial, then $M:N$ is divisorial.
2.7 Properties of $N:M$.

§3. Completely Integrally Closed Rings.
3.1 x is almost integral if and only if $x\mathfrak{a} \subseteq \mathfrak{a}$ for some fractionary ideal $\mathfrak{a}$.
3.2 The set of almost integral elements over A forms a ring $\tilde{A}$.
3.3 A is completely integrally closed if and only if $A = \mathfrak{a}:\mathfrak{a}$ for all fractionary ideals $\mathfrak{a}$.
3.4 The monoid of divisorial ideals is a group if and only if A is completely integrally closed.
3.5 If A is completely integrally closed then maximal divisorial ideals are prime ideals.
3.6 Characterization of a Krull domain in terms of divisorial ideals.
3.7 If $\mathfrak{p}, \mathfrak{q} \in \operatorname{Spec} A$ and $\mathfrak{p} \subsetneqq \mathfrak{q}$, then $A:\mathfrak{p} \subseteq \mathfrak{q}:\mathfrak{q}$.
3.8 If $\mathfrak{p} \in \operatorname{Spec} A$, then $A_\mathfrak{p} = \{x \in K^* : Ax^{-1} \cap A \not\subseteq \mathfrak{p}\} \cup \{0\}$.

Chapter II. The Divisor Class Group and Factorial Rings

18.10 $\mathrm{Pic}(\mathrm{Spec}\,A-\{\mathfrak{m}\})\to \mathrm{Cl}(A)$ is an injection.

18.11 A is factorial if and only if A is parafactorial and $\mathrm{Spec}\,A-\{\mathfrak{m}\}$ is locally factorial.

18.12 If $\hat{A}$ is parafactorial then A is parafactorial.

18.13 A complete intersection of dimension at least four is parafactorial.

18.14 Crucial lemma to prove 18.13.

18.15 A complete intersection factorial in codimension 3 is factorial.

18.16 $\mathrm{Pic}(A)\to \mathrm{Pic}(A/\mathrm{rad}\,A)$ is an injection and a bijection if A is complete.

18.17 $\mathrm{Pic}(A)\to \mathrm{Pic}(\hat{A})$ is an injection.

18.18 If $\hat{A}$ is locally factorial and $\mathrm{Spec}\,A/\mathfrak{a}$ is connected, then $\mathrm{Pic}(A/\mathfrak{a})=\mathrm{Cl}(\hat{A})$.

18.19 $\hat{A}$ is factorial if and only if $\mathrm{Pic}(A/\mathfrak{a})=0$.

18.20 Another case in which $\mathrm{Cl}(A)=\mathrm{Pic}(A/\mathfrak{a})$.

18.21 $\mathrm{Pic}(A)=\mathrm{Pic}\big(A[[T]]\big)$.

18.22 If $A_{\mathfrak{m}}[[T]]$ is factorial for all maximal ideals $\mathfrak{m}$ of A, then $\mathrm{Cl}(A)\to \mathrm{Cl}\big(A[[T]]\big)$ is a bijection.

18.23 If A is regular, then $\mathrm{Cl}(A)\to \mathrm{Cl}\big(A[[T]]\big)$ is a bijection.

§19. Completions, Formal Power Series and Danilov's Results.

19.1 If depth $A\geqslant 3$ and A is almost factorial and complete, then $A[[T]]$ is almost factorial.

19.2 Helping lemma for 19.1.

19.3 Helping lemma for 19.1.

19.4 If A is complete with depth $A\geqslant 3$, then A is factorial if and only if $A[[T]]$ is factorial.

19.5 Under the same assumption, $A[[T_1,...,T_n]]$ is factorial.

19.6 Examples of Salmon and Danilov.

19.7 Samuel's theorem. When $A[[T]]$ is not factorial.

19.8 Weierstrass preparation theorem.

19.9 Example of a completion which is not factorial.

19.10 Example.

19.11 $\big(k[x,y,z]/(x^2+y^3+z^5)\big)[[T]]$ is factorial.

19.12 Lipman's theorem.

19.13 Other cases of $k[x,y,z]/(x^r+y^s+z^t)$.

19.14 $\mathrm{Cl}\big(A[[T]]\big)\to \mathrm{Cl}\big(A[[T]]_{\mathfrak{m}A[[T]]}\big)$ is a bijection.

19.15 If A is an analytically normal local ring, then $\mathrm{Cl}\big(A[[T_1,...,T_n]]\big)$ $\to \mathrm{Cl}\big(A[[T_1,...,T_n]]_{\mathfrak{m}A[[\{T_i\}]]}\big)$ is a bijection.

19.16 If $A\to B$ is faithfully flat and B has discrete divisor class group (ddcg), then A has ddcg. If $A_{\mathfrak{m}}$ has ddcg for all $\mathfrak{m}$, then A has ddcg.

19.17 Lemma to prove 19.16.

19.18 Problem.

19.19 If A is excellent and normal and residue class field of characteristic zero, and if A has ddcg, then $S^{-1}A$ has ddcg.

19.20 A has ddcg if and only if $\mathrm{H}^1(Y,\mathcal{O}_Y)=0$.

19.21 If $A\to B$ is regular and A has ddcg, then B has ddcg.

19.22 If A is resolvable and has ddcg, then A has ddcg.

19.23 $\mathrm{H}^0(U,\tilde{a})$ and $\mathrm{H}^1(U,\tilde{a})$ are of finite type for $U=\mathrm{Spec}\,A-\{\mathfrak{m}\}$.

19.24 If depth $A\geqslant 3$ and $A_{\mathfrak{p}}$ has ddcg for $\mathfrak{p}\neq\mathfrak{m}$, then A has ddcg.

19.25 If A is locally allowable with ddcg for depth $A_{\mathfrak{p}}=2$, then A has ddcg.

19.26 If A is locally Cohen-Macaulay and if $A_{\mathfrak{p}}[[T]]$ is factorial for all $\mathfrak{p}$ with $\mathrm{ht}\,\mathfrak{p}=2$, then A has ddcg.

19.27 If A is (S_3) and (R_2), then A has ddcg.

19.28 A has ddcg if and only if $\mathrm{Cl}(^{\mathrm{sh}}A)$ is finite.

19.29 A has ddcg if and only if $\mathrm{H}^1(\mathrm{Spec}\,A-\{\mathfrak{m}\},W_\infty)=0$.

19.30 If $\dim A \geqslant 3$, depth $A = 2$, and the characteristic of residue class field is zero, then A does not have ddcg.

19.31 Example for 19.30.

19.32 $\mathbb{Q}$ algebras have ddcg if and only if (S_3) and $\mathrm{Cl}(^{\mathrm{sh}}A_p)$ is finite for $\mathrm{ht}\, p = 2$.

19.33 If $A \to B$ is regular, then A has ddcg if and only if B has ddcg.

19.34 If $A \to B$ is faithfully flat with separable extension of residue class field, then A has ddcg if and only if B has ddcg.

19.35 If A has ddcg, then $A[X_1, \ldots, X_n]$ has ddcg.

19.36 If A has ddcg, then $^{\mathrm{h}}A$ and $^{\mathrm{sh}}A$ have ddcg.

19.37 If A is excellent, then the $\mathfrak{m}$-adic completion of A has ddcg.

19.38 $\mathrm{Cl}(A) \to \mathrm{Cl}(A[[T_1, \ldots, T_n]])$ is a bijection for all n.

19.39 If A has (R_2) and is a factor of a regular ring, then $\mathrm{Cl}(A) \to \mathrm{Cl}(A[[T_1, \ldots, T_n]])$ is a bijection.

Bibliography

Most of the entries in this bibliography are from the period after 1960. The references prior to that date are included for historical reasons. Gilmer's treatise on *Multiplicative Ideal Theory* [1968] contains an extensive bibliography for the period 1934—1968 (from Krull's *Ideal theorie* to Gilmer's *Multiplicative Ideal Theory*), but it is directed more to the subject of ideal theory than to the factorization theory. Nagata (*Local Rings* [1964]) and Bourbaki (Ch. VII [1965]) both contain historical remarks, the first related to the topics presented in Nagata's text, the second a comprehensive history of the subject from its beginning. Both contain valuable bibliographies.

The following bibliography includes both references from the text and entries which may be of interest in closely related areas.

Akizuki, Y.: Theory of Local Rings. Lecture Notes. The University of Chicago 1958.

Andreotti, A., Salmon, P.: Anelli con unica decomponibilitá in fattori primi. Monatsh. Math. **61**, 97—142 (1957).

Arezzo, D., Greco, S.: Sul gruppo delle classi di ideali. Ann. Scuola Norm. Sup. Pisa, Cl. di Sc. XXI, Fase IV, 459—483 (1967).

Arezzo, D., Robbiano, L.: Sul completamento di un annello rispetto ad un ideale di tipo finito. Rend. Sem. Mat. Univ. Padova **44**, 133—154 (1970).

Armitage, J. V.: On unique factorization in algebraic function fields. J. London Math. Soc. **38**, 55—59 (1963).

Artin, Emil: Theory of Algebraic Numbers. Notes by G. Würges. Math. Inst. Göttingen. Translated by G. Stricker. Göttingen (1959).

Auslander, M., Buchsbaum, D.: Homological dimension in local rings. Trans. Amer. Math. Soc. **85**, 390—405 (1957).

Auslander, M., Buchsbaum, D.: Codimension and multiplicity. Ann. of Math. **68**, 625—656 (1958).

Auslander, M., Buchsbaum, D.: Homological dimension in noetherian rings, II. Trans. Amer. Math. Soc. **88**, 194—206 (1958).

Auslander, M., Buchsbaum, D.: Unique factorization in regular local rings. Proc. Nat. Acad. Sci. U. S. A. **45**, 733—734 (1959).

Auslander, M. Buchsbaum, D.: On ramification theory in noetherian rings. Amer. J. Math. **81**, 749—765 (1959).

Auslander, M., Buchsbaum, D.: Invariant factors and two criteria for projectivity of modules. Trans. Amer. Math. Soc. **104**, 516—522 (1962).

Bass, H.: On the ubiquity of Gorenstein rings. Math. Z. **82**, 8—28 (1963).

Bass, H.: Algebraic K-theory. New York-Amsterdam: Benjamin 1968.

Beck, I.: Injective modules oves a Krull domain. J. Algebra **17**, 116—131 (1971).

Bertin, M.-J.: Anneaux d'invariants d'anneaux de polynomes, en caractéristique *p*. C. R. Acad. Sci. Paris. **264**, 653—656 (1967).

Bertin, M.-J.: Sous-anneaux d'invariants d'anneaux de polynomes. C. R. Acad. Sci. Paris. **260**, 5655—5658 (1965).

Bertin, M.-J.: Anneau des invariants du groupe alterné, en caractéristique 2. Bull. Sci. Math. **94**, 65—72 (1970).

Blankinship, W. A.: A new version of the Euclidean algorithm. Math. Monthly **70**, 742—745 (1963).

Borevich, Z. I., Shafarevich, I. R.: Number Theory. Translated from Russian by Newcomb Greenleaf. New York and London: Academic Press 1966.

Bourbaki, N.: Algèbre Commutative. Ch. I, II. Paris: Hermann 1961. Ch. III, IV. Paris: Hermann 1961. Ch. V, VI. Paris: Hermann 1964. Ch. VII. Paris: Hermann 1965.

Boutot, J.-F.: Groupe de Picard local d'un anneau hensélien. C. R. Acad. Sci. Paris. Sér. A **272**, 1248—1250 (1971).

Brieskorn, E.: Beispiele zur Differential-Topologie von Singularitäten. Invent. Math. **2**, 1—14 (1966).

Brieskorn, E.: Rationale Singularitäten komplexer Flächen. Invent. Math. **4**, 336—358 (1968).

Buchsbaum, D. A.: Some remarks on factorization in power series rings. J. Math. Mech. **10**, 749—753 (1961).

Cartan, E., Eilenberg, S.: Homological Algebra. Princeton: Princeton University Press 1956.

Cashwell, E. D., Everett, C. J.: The ring of number theoretic functions. Pacific J. Math. **9**, 975—985 (1959).

Cashwell, E. D., Everett, C. J.: Formal power series. Pacific J. Math. **13**, 45—69 (1963).

Claborn, L.: The dimension of $A[X]$. Duke Math. J. **32**, 233—236 (1965).

Claborn, L.: Dedekind domains and rings of quotients. Pacific J. Math. **15**, 59—64 (1965).

Claborn, L.: Dedekind domains: Overrings and semi-prime elements. Pacific J. Math. **15**, 799—804 (1965).

Claborn, L.: Note generalizing a result of Samuel's. Pacific J. Math. **15**, 805—808 (1965).

Claborn, L.: Every abelian group is a class group. Pacific J. Math. **18**, 219—222 (1966).

Claborn, L.: A note on the class group. Pacific J. Math. **18**, 223—225 (1966).

Claborn, L.: A generalized approximation theorem for Dedekind domains. Proc. Amer. Math. Soc. **18**, 378—380 (1967).

Claborn, L.: Specified relations in the ideal group (I) Michigan Math. J. **15**, 249—255 (1968).

Claborn, L., Fossum, R.: Higher rank class groups. Bull. Amer. Math. Soc. **73**, 233—237 (1967).

Claborn, L., Fossum, R.: Generalizations of the notion of class group. Illinois J. Math. **12**, 228—253 (1968).

Claborn, L., Fossum, R.: Class groups of n-noetherian rings. J. Algebra. **10**, 263—285 (1968).

Cohen, I. S.: On the structure and ideal theory of complete local rings. Trans. Amer. Math. Soc. **59**, 54—106 (1946).

Cohen, I. S., Seidenberg, A.: Prime ideals and integral dependence. Bull. Amer. Math. Soc. **52**, 252—261 (1946).

Cohn, P. M.: Bézout rings and their subrings. Proc. Cambridge Philos. Soc. **64**, 251—264 (1968).

Cunnea, W. M.: Unique factorization in algebraic function fields. Illinois J. Math. **8**, 425—438 (1964).

Danilov, V. I.: The group of ideal classes of a completed ring. Math. USSR-Sb. **6**, 493—500 (1968). (Mat. Sb. **77** (119), 533—541 (1968).)

Danilov, V. I.: On a conjecture of Samuel. Math. USSR-Sb. **10**, 127—137 (1970). (Mat. Sb. **81** (123), 132—144 (1970).)

Danilov, V. I.: Rings with a discrete group of divisor classes. Math. USSR-Sb. **12**, 368—386 (1970). (Mat. Sb. **83** (125), 372—389 (1970).)

Danilov, V. I.: Rings with a discrete group of divisor classes. Mat. Sb. **88** (130), 229—237 (1972).

Davis, Edward D.: Rings of algebraic numbers and functions. Math. Nachr. **29**, 1—7 (1965).

Davis, Edward D.: Ideals of the principal class, R-sequences and a certain monoidal transformation. Pacific J. Math. **20**, 197—205 (1967).

Deckard, D., Durst, L. K.: Unique factorization in power series rings and semigroups. Pacific J. Math. **16**, 239—242 (1966).

Deuring, M.: Algebren. 2. Auflage. Ergebnisse der Mathematik und ihrer Grenzgebiete, Bd. 41, Berlin-Heidelberg-New York: Springer 1968.

Dieudonné, J.: Topics in Local Algebra. Notes by Mario Borelli. Notre Dame. Math. Lect. No. 10. Notre Dame, Indiana: University of Notre Dame Press 1967.

Eakin, P. M., Heinzer, W. J.: Some open questions on minimal primes of Krull domains. Canad. J. Math. **20**, 1261—1264 (1968).

Eakin, P. M., Heinzer, W. J.: Non finiteness in finite dimensional Krull domains. J. Algebra **14**, 333—340 (1970).

Fiorentini, M.: Esempi di anelli di Cohen-Macaulay semifattoriali che non sons di Gorenstein. Accad. nazionale dei Lincei. Rendiconti **50**, 524—529 (1971).

Fossum, R., Griffith, P., Reiten, I.: The homological algebra of trivial extensions of abelian categories and applications to ring theory. (To appear.)

Foxby, H.-J.: On Gorenstein modules and related modules. Mimeograph: Københavns Universitet Matematisk Institut. Preprint series No. 14. 1971.

Garfinkel, G.: Generic splitting algebras for Pic. Pacific J. Math. **85**, 369—380 (1970).

Geramita, A. V., Roberts, L. G.: Algebraic vector bundles on projective space. Invent. Math. **10**, 298—304 (1970).

Gilmer, R.: Some applications of the Hilfssatz von Dedekind-Mertens. Math. Scand. **20**, 240—244 (1967).

Gilmer, R.: Multiplicative Ideal Theory, Parts I and II. Queen's Papers on Pure and Applied Mathematics. No. 12, Kingston, Ontario: Queen's University Press 1968.

Gilmer, R.: Power series rings over a Krull domain. Pacific J. Math. **29**, 543—549 (1969).

Gilmer, R.: An embedding theorem for HCF rings. Proc. Cambridge Philos. Soc. **68**, 583—587 (1970).

Gilmer, R., Heinzer, W.: Rings of formal power series over a Krull domain. Math. Z. **106**, 379—387 (1968).

Giraud, J.: Groupe de Picard, anneaux factoriels. Sem. Bourbaki 15 (1962/63) exp. no. 248 (1964).

Goldman, O.: On a special class of Dedekind domains. Topology **3**, 113—118 (1964).

Greco, S.: Sull' integrità e la fattorialità dei complementi m-adici. Rend. Sem. Mat. Univ. Padova **36**, 50—65 (1966).

Greco, S.: Sugli ideali frazionari invertibili. Rend. Sem. Mat. Univ. Padova **36**, 315—333 (1966).

Greco, S.: Anelli di Gorenstein. Sem. Inst. Mat. Univ. Genova 1969.

Greco, S.: Henselization of a ring with respect to an ideal. Trans. Amer. Math. Soc. **144**, 43—65 (1969).

Greco, S.: Sugli omorfismi piatti e non ramificati. Matematiche, Catania XXIV. fasc. **2**, 392—415 (1969).

Greco, S., Salmon, P.: Anelli di Macaulay. Pubbl. Inst. Mat. Univ. di Genova 1965.

Greco, S., Salmon, P.: Topics in $\mathfrak{m}$-adic Topologies. Ergebnisse der Mathematik und ihrer Grenzgebiete, Bd. 58, Berlin-Heidelberg-New York: Springer 1971.

Grothendieck, A.: Cohomologie locale des faisceaux coherents et theoremes de Lefschetz locaux et globaux. Séminaire de Géométrie Algébrique (SGA) 1962. IHES. fase I and II 1963.

Grothendieck, A.: Local Cohomology. Lecture Notes in Mathematics. **41**. Berlin-Heidelberg-New York: Springer 1967.

Grothendieck, A., Dieudonné, J.: Eléments de Géométrie Algébrique. I. IHES Publ. Math. No. 4, 1960. IV. (Premiere partie), No. 20, 1964. IV. (Seconde partie), No. 24, 1965. IV. (Troisième partie), No. 28, 1966. IV. (Quatrième partie), No. 32, 1967. (Refer to as EGA, IV, etc.)

Hallier, N.: Quelques propriétés arithmétiques des dérivations. C. R. Acad. Sci. Paris. **258**, 6041—6044 (1964).

Hallier, N.: Utilisation des groupes de cohomologie. C. R. Acad. Sci. Paris. **261**, 3922—3924 (1965).

Halmos, P.: How to write mathematics. Enseignement Math. **16**, 123—152 (1970).

Heinzer, W.: On Krull overrings of an affine ring. Pacific J. Math. **29**, 145—149 (1969).

Heinzer, W.: On Krull overrings of a noetherian domain. Proc. Amer. Math. Soc. **22**, 217—222 (1969).

Heinzer, W.: Quotient overrings of integral domains. Mathematika **17**, 139—148 (1970).

Hironaka, H.: Resolution of singularities of algebraic varieties over fields of characteristic zero. I, II. Ann. of Math. (2) **79**, 109—203, 205—326 (1964).

Hironaka, H.: On the characteristic v^* and τ^* of singularities. J. Math. Kyoto Univ. **7**, 19—43 (1967).

Hochster, M.: Homogeneous coordinate rings of Schubert varieties are Cohen-Macaulay. *(To appear.)*

Hochster, M., Eagon, J. A.: A class of perfect determinantal ideals. Bull. Amer. Math. Soc. **76**, 1026—1029 (1970).

Hochster, M., Eagon, J. A.: Cohen-Macaulay rings, invariant theory, and the generic perfection of determinantal loci. Amer. J. Math. **93**, 1020—1056 (1971).

Jacobson, N.: Lectures in Abstract Algebra. Vol. III. Princeton: Van Nostrand 1964.

Jaffard, P.: Les System d'Ideaux. Paris: Dunod 1960.

Kaplansky, I.: Modules over Dedekind rings and valuation rings. Trans. Amer. Math. Soc. **72**, 327—340 (1952).

Kaplansky, I.: Commutative Rings. Allyn and Bacon: Boston 1970.

Koester, A.: Lokale Darstellbarkeit eindimensionaler analytischer Mengen in zweidimensionalen normalen komplexen Räumen durch eine holomorphe Funktion. Schriftreihe des Math. Inst. der Univ. Münster. Heft 35. 1967.

Krull, W.: Allgemeine Bewertungstheorie. J. Reine Angew. Math. **167**, 160—196 (1931).

Krull, W.: Idealtheorie. Ergebnisse der Mathematik und ihrer Grenzgebiete. Berlin: Springer 1935. (Band 46) 2., ergänzte Aufl. Berlin-Heidelberg-New York: Springer 1968.

Krull, W.: Beiträge zur Arithmetik kommutativer Integritätsbereiche. Math. Z. **41**, 545—577 (1936).

Krull, W.: II. Math. Z. **41**, 665—679 (1936).

Krull, W.: III. Math. Z. **42**, 745—766 (1937).

Krull, W.: IV. Math. Z. **42**, 767—773 (1937).

Krull, W.: V. Math. Z. **43**, 768—782 (1938).

Krull, W.: Zur Arithmetik der endlichen diskreten Hauptordnungen. J. Reine Angew. Math. **189**, 118—128 (1951).

Krull, W.: Jacobsonsche Ringe, Hilbertscher Nullstellensatz, Dimensionstheorie. Math. Z. **54**, 354—387 (1951).

Krull, W.: Zur Theorie der kommutativen Integritätsbereiche. J. Reine Angew. Math. **192**, 230—252 (1953).

Krull, W.: Einbettungsfreie fast-noethershe Ringe und ihre Oberringe. Math. Nachr. **21**, 319—338 (1960).

Lafon, J. P.: Anneaux Henséliens. Bull. Soc. Math. France **91**, 77—107 (1963).

Leedham-Green, C. R.: The class group of Dedekind domains. Trans. Amer. Math. Soc. **163**, 1—8 (1972).

LeVeque, W. J.: Topics in Number Theory. Reading, Mass.: Addison-Wesley 1956.

Lipman, J.: Rational Singularities with applications to algebraic surfaces and unique factorization. IHES Publ. Math. No. 36, 195—279 (1969).

Lu, Chin-Pi.: On the unique factorization theorem in the ring of number theoretic functions. Illinois J. Math. **9**, 40—46 (1965).

Lu, Chin-Pi.: On the concordance of local rings and unique factorization. Illinois J. Math. **11**, 291—296 (1967).

Lu, Chin-Pi.: A generalization of Mori's theorem. Bull. Amer. Math. Soc. **31**, 373—375 (1972).

MacRae, R.: On the homological dimension of certain ideals. Proc. Amer. Math. Soc. **14**, 746—750 (1963).

MacRae, R.: On an application of the Fitting invariants. J. Algebra **2**, 153—169 (1965).

Matsumura, H.: Commutative Algebra. New York: Benjamin 1970.

Mattuck, A.: Complete ideals and monoidal transforms. Proc. Amer. Math. Soc. **26**, 555—560 (1970).

Mertens, F.: Über einen algebraischen Satz. S.-B. Akad. Wiss. Wien (2a) **101**, 1560—1566 (1892).

Micali, A.: Sur les algèbres universelles. Ann. Inst. Fourier **14**, 33—88 (1964).

Micali, A.: Algèbres intègres et sans torsion. Bull. Soc. Math. France **94**, 5—14 (1966).

Micali, A., Salmon, P., Samuel, P.: Intégrité et factorialité des algèbres symétriques. Atas quarto Coloquio Brasil. Mat. Posos de Caldas 1963, 61—76 (1965).

Mori, Y.: On the integral closure of an integral domain. Mem. Coll. Sci. Univ. Kyoto **27**, 249—256 (1952—53).

Mori, Y.: Errata, Ibid. **28**, 327—328 (1953—54).

Mori, Y.: On the integral closure of an integral domain. Bull. Kyoto Gakugie Univ. Ser. B. **7**, 19—30 (1955).

Mori, Y.: On the fundamental theorem of regular local rings. Bull. Kyoto Gakugie Univ. Ser. B. **15**, 17—22 (1959).

Motzkin, T.: The Euclidean algorithm. Bull. Amer. Math. Soc. **55**, 1142—1146 (1949).

Mumford, D.: Introduction to Algebraic Geometry. Preliminary Edition. Harvard, Cambridge, Mass. 1966.

Mumford, D.: Lectures on Curves on an Algebraic Surface. Ann. of Math. Studies No. 59. Princeton: Princeton Univ. Press 1966.

Mumford, D.: Abelian Varieties. Bombay: Oxford University Press for Tata Institute of Fundamental Research 1970.

Murthy, M.-Pavaman: A note on factorial rings. Arch. Math. **15**, 418—420 (1964).

Murthy, M.-Pavaman: Projective modules over a class of polynomial rings. Math. Z. **88**, 184—189 (1965).

Murthy, M.-Pavaman: Vector bundles over affine surfaces birationally equivalent to a ruled surface. Ann. of Math. **89**, 242—253 (1969).

Nagata, M.: On the derived normal rings of noetherian integral domains. Mem. Coll. Sci. Univ. Kyoto **29**, 293—303 (1955).

Nagata, M.: A general theory of algebraic geometry over Dedekind domains. I. Amer. J. Math. **78**, 78—116 (1956). II. Amer. J. Math. **80**, 382—420 (1958). III. Amer. J. Math. **81**, 401—435 (1959).

Nagata, M.: A remark on the unique factorization theorem. J. Math. Soc. Japan **9**, 143—145 (1957).

Nagata, M.: Local Rings. Interscience Tracts in Pure and Applied Mathematics. No. 13, New York: Interscience 1962.

Nagata, M.: Lectures on the fourteenth problem of Hilbert. Tata Inst. of Fundamental Research Lectures on Math. No. 31. Bombay: Tata Institute 1965.

Nicolas, A.-M.: Modules factoriels. Bull. Sci. Math. **95**, 33—52 (1971).

Nishimura, H.: On the unique factorization theorem for formal power series. J. Math. Kyoto Univ. **7**, 151—160 (1967).

Northcott, D.G.: Some remarks on the theory of ideals defined by matrices. Quart. J. Math. Oxford Ser. (2) **14**, 193—204 (1963).

Peskine, C., Szpiro, L.: Notes sur un air de H. Bass. Mimeographed notes. Brandeis University 1968.

Pirtle, E.M.: A generalization of the class group. Notices Amer. Math. Soc. **17**, 762. 677—13—1 (1970).

Pirtle, E.M.: Generalized factorial rings. Notices Amer. Math. Soc. **17**, 807 (1970).

Prill, D.: Local classification of quotients of complex manifolds by discontinuous groups. Duke Math. J. **34**, 375—386 (1967).

Prill, D.: The divisor class groups of some rings of holomorphic functions. Math. Z. **121**, 58—80 (1971).

Ramis, J.-P.: Factorialité des anneaux de séries formelles et de séries convergentes sur les espaces vectoriels normes. C.R. Acad. Sci. Paris. Sér. A. **262**, 904—906 (1966).

Ramis, J.-P.: Factorialité des anneaux de polynômes de séries formelles et de séries convergentes. Théorème de préparation en géométrie différentiable banachique. Sem. Analyse P. Lelong 7 (1966/67), No. 3, 14 p (1967).

Raynaud, M.: Anneaux Locaux Hensèliens. Lecture Notes in Mathematics No. 169. Berlin-Heidelberg-New York: Springer 1970.

Richman, F.: Generalized quotient rings. Proc. Amer. Math. Soc. **16**, 794—799 (1965).

Robbiano, Lorenzo: Sulla fattorialita di certi anelli henseliani. Boll. Un. Mat. Ital. IV. Ser. **2**, 667—681 (1969).

Rosenberg, A., Zelinsky, D.: Automorphisms of separable algebras. Pacific J. Math. **11**, 1109—1117 (1961).

Sally, Judith: A note on integral closure. Typescript 1971 (To appear).

Salmon, P.: Sur les séries formelles restreintes. Bull. Soc. Math. France **92**, 385—410 (1964).

Salmon, P.: Serie convergenti su un corpo non archimedeo con applicazione ai fasci analitici. Ann. Mat. Pura Appl. Ser. IV, **LXV**, 113—125 (1964).

Salmon, P.: Su un problema posto da P. Samuel. Atti Accad. Naz. Lincei Rend. Cl. Sci. Fis. Mat. Natur. (8) **40**, 801—803 (1966).

Salmon, P.: Sulla fattorialità delle algebre graduate e degli anelli locali. Rend. Sem. Mat. Univ. Padova **41**, 119—138 (1968).

Salmon, P.: Singolaritá e gruppo di Picard. Inst. Naz. di Alta Mat., Symposia Matematica II, 341—345 (1968).

Samuel, P.: On unique factorization domains. Illinois J. Math. **5**, 1—17 (1961).

Samuel, P.: Les anneaux factoriels. Atti della 2ª Riunione del Groupement de Mathématiciens d'expression latine Firenze, 26—30. Sept. 1961. Bologna. 1—3 ott. 1961, 64—65.

Samuel, P.: Sur les anneaux factoriels. Bull. Soc. Math. France **89**, 155—173 (1961).

Samuel, P.: Un exemple d'anneau factorial. Bol. Soc. Mat. São Paulo (1962).

Samuel, P.: Anneaux gradués factoriels et modules réflexifs. Bull. Soc. Math. France **92**, 237—249 (1964).

Samuel, P.: Classes de diviseurs et dérivées logarithmiques. Topology **3**, Suppl. **1**, 81—96 (1964).

Samuel, P.: Anneaux Factoriels. (red. A. Micali), Bol. Soc. Mat., São Paulo (1964).

Samuel, P.: Lectures on Unique Factorization Domains. Notes by P. Murthy. Tata Inst. for Fundamental Research, No. 30. Bombay 1964.

Samuel, P.: Modules réflexifs et anneaux factoriels. Les Tendances Géométriques en algèbre et théorie des nombres No. 143. Colloques Internationaux Centre National de la Recherche Scientifique. 1966.

Samuel, P.: Modules réflexifs et anneaux factoriels. Algèbre Théorie Nombres, Sem. P. Dubreil, M.-L. Dubreil-Jacobin, L. Lesieur et C. Pisot 17 (1963/64), No. 8, 4 p (1967).

Samuel, P.: On a construction due to P. M. Cohn. Proc. Cambridge Philos. Soc. **64**, 249—250 (1968).

Samuel, P.: Unique factorization. Amer. Math. Monthly **75**, 945—952 (1968).

Samuel, P.: About Euclidean rings. J. Algebra **19**, 282—301 (1971).

Scheja, G.: Über die Bettizahlen lokaler Ringe. Math. Ann. **155**, 155—172 (1964).

Scheja, G.: Über Primfaktorzerlegung in zweidimensionalen lokalen Ringen. Math. Ann. **159**, 252—258 (1965).

Scheja, G.: Einige Beispiele faktorieller lokaler Ringe. Math. Ann. **172**, 124—134 (1967).

Scheja, G., Wiebe, H.: Uber die Divisorenklassengruppen lokaler Ringe auf zweidimensionalen normalen Hyperflächen. Math. Ann. **172**, 229—237 (1967).

Selmer, E.S.: The Diophantine equation $ax^3 + by^3 + cz^3 = 0$. Acta Math. **85**, 203—362 (1951).

Serre, J.-P.: Sur la dimension homologique des anneaux et des modules noethériens. Proc. Int. Symp. on Algebraic Number Theory. Tokyo 175—189 (1956).

Serre, J.-P.: Corps Locaux. Paris: Hermann 1962.

Singh, B.: Unique factorization in power series rings. Invent. Math. **3**, 348—355 (1967).

Singh, B.: On a conjecture of Samuel. Math. Z. **105**, 157—159 (1968).

Singh, B.: Invariants of finite groups acting on local unique factorization domains. J. Indian Math. Soc. **34**, 31—38 (1970).

Storch, U.: Fastfaktorielle Ringe. (Schriftenreihe Math. Inst. Univ. Münster. Heft 36.) Münster (Westf.): Max Kramer 1967.

Storch, U.: Über die Divisorklassengruppen normaler komplexanalytischer Algebren. Math. Ann. **183**, 93—104 (1969).

Strooker, J.R.: A remark on Artin-Van der Waerden equivalence of ideals. Math. Z. **93**, 241—242 (1966).

Sweedler, M.E.: Cohomology of algebras over Hopf algebras. Trans. Amer. Math. Soc. **133**, 205—239 (1968).

Van der Waerden, B.L.: Moderne Algebra, II. Berlin: Springer 1931.

Waterhouse, W. C.: Divisor classes in pseudo Galois extensions. Pacific J. Math.
    36, 541—548 (1971).
Yuan, Shuen: On logarithmic derivatives. Bull. Soc. Math. France **96**, 41—52
    (1968).
Zariski, O.: Pencils on an algebraic variety and a new proof of a theorem of Bertini.
    Trans. Amer. Math. Soc. **50**, 48—70 (1941).
Zariski, O.: The concept of a simple point of an abstract algebraic variety. Trans.
    Amer. Math. Soc. **62**, 1—52 (1947).
Zariski, O., Samuel, P.: Commutative Algebra. Vol. I. Princeton: Van Nostrand
    1958. Vol. II. Princeton: Van Nostrand 1960.
Zinn-Justin, Nicole (nee Hallier): Derivations dans les corps et anneaux de cha-
    ractéristique p. Bull. Soc. Math. France, Suppl., Mém. No. **10**, 79 p (1967).
Zinn-Justin, Nicole (nee Hallier): Descente p-radicelle et factorialité. Colloque
    d'algèbre, 6—7 mai 1967, Paris: Secrètariat mathématique 1968.

Index

Ergebnisse der Mathematik und ihrer Grenzgebiete

Prices are subject to change without notice